GOOGLE CLASSROOM

THE BEST TOOL FOR DISTANCE LEARNING. HOW TO OPTIMIZE THE TASKS, STRENGTHEN COLLABORATION, AND FACILITATE COMMUNICATION TO MAKE TEACHING MORE PRODUCTIVE

Author: Alex Abdow

Table of Contents

—

Introduction

Education has gone digital for quite some time. Educators use digital platforms to successfully exchange information with the students, discuss tasks, plan lessons, and check plagiarism in papers. However, there have been numerous free software tools available online, which still involves some added costs in using them. In this book, the whole focus will be on the optimal use of this online and virtual learning medium. It will explain the maximum benefits that individuals can extract from it. Utilizing Google Classroom could save time for students and teachers, as well as make learning more efficient. High technology, including comprehensible design, makes it much easier for teachers to develop. Besides, they can focus on the task and also the main purpose of the class rather than focusing on solving trivial problems like printing and distributing assignments and explaining content to every individual student. This can dramatically reduce the need for funds to further improve student learning outcomes.

In the changing world today, students have to finish school with such a range of technical skills and abilities that will help and guide them to attain success in the job market around the world. There is a powerful fantastic platform for students to access those skills and it is Google Classroom. Within global learning societies, there are debates every day about how educators can better prepare the students to progress in the educational and work fields. Understanding what skills the students will mostly require in the coming years is the difficult part of such discussions. Educators are forced to find out how to train the students for such a future not yet in existence. In order to do that, they need not only to decide how to use the resources and expertise we currently possess but also to anticipate the skills today's students would need in the future. Empowering children with the essential and versatile expertise should be a must so that they can have the greatest benefit of adjusting to whatever the environment looks like. Here are some advantages of this method, when used during the Classroom to support educators to prepare the students for all the developments of the modern age.

An Exposure towards Online Learning Methods and Platforms

A lot of colleges today expect students to learn at least some online courses during the degree process. In reality, when you get a Master's degree in education, those of your online coursework might be valid. Sadly, many of the students never had that online education experience. That is why, at an early stage, you need to ensure absolutely that your students have as much support as possible. Google Classroom would be a simple way for students to assist with this process since it is extremely user-friendly.

Simple Material Access

Google Classroom allows students access to materials regardless of where they are, as everything is shared online. The missing days of the rubrics and worksheets are gone. When required, students who have not been available may easily manage classroom resources from home — this will also help to save you and the students a great deal of stress over the long term.

The Differentiation

Google Classroom has become a perfect resource for differentiation, as you would set up several different classrooms. When you focus on a topic in the Classroom or have teams that focus on two distinct levels, essentially build two different types of classes with that subject. This means you can really reach out to those who could not create good work even after doing a lot of hard work despite trying to make them feel dumb or wrong.

It can enable you to deliver assignments on an even more individualized level, and can also help to reach out to some students. You can also break people into groups you think those who can work much better together. Google Classroom would be a perfect, versatile way to make sure each student gets what they want and need, so when you see appropriate, you could conveniently erase and recreate lessons.

Less Usage of Paper

Google Classroom may practically rid itself of paper usage when it is used to administer a whole class. All classwork could be done online as long as the students possess internet access. That indicates no copies for the district and therefore less cash.

No Possibility of Lost Work and Progress

Students could not miss their research unless they have it physically during their presence. Because they usually operate on Google Drive, it is all immediately saved; therefore, excuses are dwindling. Students will encounter more organizational performance from a little short lesson about how to effectively use such online resources.

The Engagement

This has been shown time after time that technology engages students. Google Classroom could encourage the student's involvement in the learning process and also makes them active learners. For example, if you already have students answering questions during Classroom, other students will comment on those answers and expand analysis for both participants.

Overall, it is certainly worth utilizing Google Classroom. This will definitely save time and money, which can make you train your students better for the potential future. The tools offered by Google regarding education through the transformation of a Classroom would be very beneficial for teachers and administrators seeking to best knowing the emerging teaching methods and also how academics around the globe are educating the students. This is rich with information and references from teachers and classrooms all over the world for perspectives. It has been quite interesting and informative to see it and understand the full results of education realized leaders' several years of studies and viewpoints. The aim is to expand the awareness of what the students could look like in the process of learning or work as well as the vision of a Classroom offers an excellent baseline for teachers and administrators.

First Steps: Free Registration on Google Classroom

Getting Started With Google Classroom

While classrooms become more paperless, teachers need to find solutions for homework sharing, classroom management, student communication, and so on.

More and more teachers are coming to Google Classroom for smooth virtual classes that focus less on technology and more on teaching. You don't need to be a tech professional to lead this department.

Before you can make Google Classroom a part of your teaching experience, you need to make sure that you download the Google Apps for Education. This is a pretty simple process to go through and have worked for you, but there are a few steps that you will need to work on first.

First, there are a few rules for who can get on the Google Classroom app. Any student, parent, or alumni group that has been registered as a 501(c) (3) can get on the Google Classroom app. Also, any accredited or non-profit K through 12 learning institutions can use this as well. You may want to consider talking to your administrator to see if this is a service that is already offered through your school or not.

Once you have checked in on this part, you will need to go through the following steps to sign up for the app:

- Start by going to the sign-up form for the Apps for Education. You will be able to find it at https://www.google.com/a/signup/?enterprise_prod uct=GOOGLE.EDU#

- Once you are on this page, you can fill the form in and then click on the Next button.

- From here, you will provide your institution's domain. If this isn't available, you can go through and purchase a new domain to help you get this started.

- Then click on Next and provide the rest of the information that is needed so that you can become an admin.

- Make sure to read through the Agreement before accepting and finishing the sign-up.

You may want to consider doing this a little bit early, like before school starts, because it can take up to two weeks before Google is done reviewing the application. After this time frame, you are going to receive your acceptance form when the application is successful. You will now be able to verify your own domain as well as add addresses, mail integration, apps, and contacts to start using this. Keep in mind that you will not need to do this for each class, but you only need to go through this and do it one time. Once you are in, you will be able to assign different classrooms and have more than one in place under the same account.

From here, you will need to work on downloading the Google Classroom App. You will be able to go to the App Store and get this one just by looking for Google Classroom, or you can just visit Google Play and find it as well. Once you have downloaded this particular app, you will need to choose to sign in as the teacher. It is also possible to download an extension or a bookmark-like app that will work on Google Chrome. You just need to visit the Chrome Store to install all of this.

There are ways you can install the Google Classroom App so that you can use it for your own needs. If you are considered the administrator, you will be able to install this extension by using the following steps:

- Visit the Google Admin Console

- From here you will click on Chrome Management and then User Settings

- Then you will be able to select the organizational unit that you would like to work with from here.

- Now you can click on Apps & Extension before clicking on Force Installed Apps & Extension. When you get to this point, click on Manage Force Installed App.

Inside this tab, you will click on Chrome Web Store and then look for the series of letters that don't seem to make sense and click on those. If you can't find that, you can click on Share to Classroom.
Now click the Add button that is right next to your extension before hitting save.
These steps are only if you are going to be the administrator. If you are using the Google Classroom through your educational institution, you will not need to go through the steps above because someone else will go through and do it. The steps that you will need to go through as a teacher, and that your students will need to go through as well, include:

- Visit g.co/share to classroom

- When you get here, you can click on Add to Chrome and then click on the Add button

- From here, you can click on the icon that is next to the extension. You will need to make sure that you are signed in with Google before you do all of this.

And that is all that you will need to do to make sure that Google Classroom is downloaded for you and that you can use it for your classroom and your teaching experience. Since this is such a popular platform to work with, it is likely that your school already has this available for you to use for your classes, or they will be interested in learning more about it.

Now your class is ready! At least it is there, and anyone can access it.

How to Create a Course on Google Classroom

Brief Instructions for Creating an Online Course in Google Classroom

We will now get acquainted with the basic elements of Google Classroom on the example of the "My course in Google Class" remote training.

When creating and organizing a course, you will have three main tabs available: Ribbon, Tasks, and Users. At first, you will see only two tabs:

- Ribbon

- Users.

You will need to add the "Tasks" tab to the course.

For a start, let us understand what a ribbon is. This is the place where current information on the course is collected and displayed: training materials, announcements, tasks, and user comments are visible.

The Task tab allows you to add training materials to the course and distribute tasks according to topics and in the required sequence.

In the Users section, there will be a list of trainees who have joined the course. This can be done by code or by adding manually. The course code can be found by clicking on the gear image.

Distance course in Google Classroom

Ribbon: information about what is happening in the course. The ribbon displays what is happening in the course in the sequence in which the teacher adds information to the course:

- Teacher announcements.

- Information about training materials for students.

- Information about the tasks for students.

- Announcements from the students themselves. This requires additional settings.

Ribbon in Google Classroom

When creating Ads, the teacher can add different materials in addition to the text, attach a file (it needs to be downloaded from a computer), add a file from Google Drive, publish a link to a video from YouTube or give a link to an external site. Students have the opportunity to view the Announcements and comment on this. If you need to add an advertisement, then you should use "Add a new entry."

Adding to the Google Classroom

It should be noted that all the downloaded material in the course ribbon in the new entry section is placed in the course folder on Google Drive. You can see the folder in the "Tasks" tab.

If you do not want your students to comment on your recording, you can disable/enable this in the Course Settings section. You will find this in the upper right corner of the course page where the gear is. There you can enable students to leave entries in the course feed or prevent them from doing so by clicking on disable.

To create the "Tasks" tab on your training course, click on the "?" in the lower left.

This is a new tab in Google Classroom. In the "Tasks" tab you can:

- Create tasks and questions, and group them by topic.

- Add educational materials of various types and combine them by topic.

- Organize the topics and materials in them; if the material does not have a topic, it is located at the top of the page.

Tasks for students can be of various types. The teacher can attach as a task any document located on a PC or Google Drive, and give a link to a video. You can also offer to perform practical work or test work in the form of a test or add a question that both teachers and other students will be able to comment on. For this, you need certain settings.

Currently, it is possible to create tasks using Google Forms. Google Forms is very versatile and with its help, it is easy to create tests with a choice of one or several answers, open tasks, and create tasks using pictures and videos, etc.

Tasks can have a set deadline. After students complete the assignments, information about this is automatically sent to the teacher. To view past assignments, the teacher goes to the ribbon section and then clicks All Tasks.

The teacher will receive information about the submitted / non-submitted work.

The teacher can check students' assignments, set grades, and comment on student answers.

As a rule, after creating a course the author creates a landing page in which he briefly presents a description of the course — the program, the start and end date of the course, the rules of work and requirements for students, and a link to the registration form.

The author of the course sends the course code to all those registered on the landing page, then students themselves are added to the course or you can manually invite students to the course. You can do this in the "Users" tab where you can see the course code and invite students by name or email address.

We remind you that students must have a Google account for classes in a training course organized on the Google Classroom platform!

How to Create a Class

As a first step, we need to access Drive or Gmail to authenticate our access to Google services. Then we have to click on the Classroom icon.

Google Classroom will be located in the top right-hand menu in the form of squares, usually in the "more" section.

Click the icon and it will open Google Classroom, there will be a window where you can enter the title of the course and be able to create it.

Create or join your first class!

You see the courses or existing classes. Create a new course via the " + " icon, by entering the fields "Class name", "Section", "Subject" etc. (the only essential is the name).

Create class

Class name (required)

Section

Subject

Room

Cancel   Create

Once confirmed with "Create", another window will open divided into four sections: "Stream", "Class works", "People" and "Grades." This is our dashboard; let's see what each of these tabs is for:

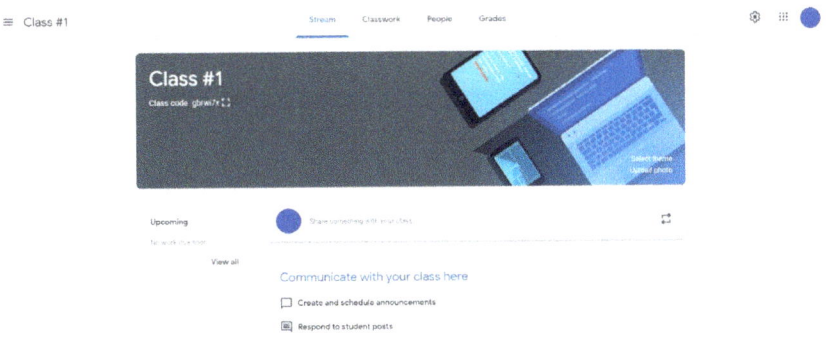

Stream

Stream is a useful section for communication between participants. It is dedicated to create and program ads, respond to posts published by students, the typical event of a normal Social network, through a bulletin board shared with those who are part of the project. You can insert messages in which you can attach: materials from Google Drive, links, files in local memories, or YouTube videos. The message is entered by clicking on the "Publish" button.

Communicate with your class here

💬 Create and schedule announcements

🗐 Respond to student posts

Course Configuration and Personalization

By clicking on the gear icon at the top right you can access the course configuration.

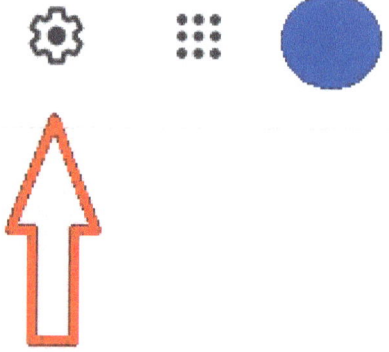

In the first part of the screen, there is general information (Class name, section, subject, etc.). Scrolling down you see other options, such as the one to decide if students can enter posts and comments (by default is active). Another important choice concerns the calculation of the vote total (none, total points, weighted by category), with the ability to show the rating to students and manage the rating categories if you opt for the weighted vote. The following shows the case of "total points."

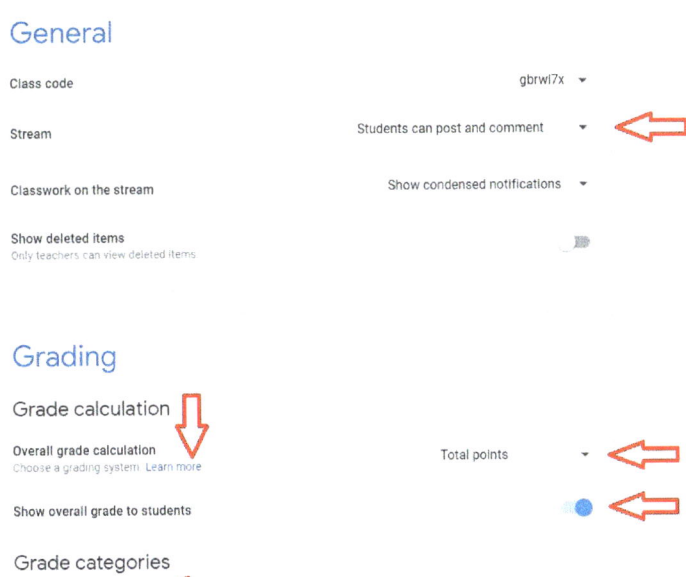

General

Class code	gbrwi7x ▾
Stream	Students can post and comment ▾ ⟸
Classwork on the stream	Show condensed notifications ▾
Show deleted items	
Only teachers can view deleted items	

Grading

Grade calculation ⬆

Overall grade calculation — Total points ▾ ⟸
Choose a grading system Learn more

Show overall grade to students — ⟸

Grade categories

Add grade category ⟸

How to assign a task

Let's see how you assign a task with a quiz (this feature rests on one of the other applications of the Suite, i.e. "modules.") Let us then return to the tab "Classwork" and click on "Create", and then on "Quiz assignment."

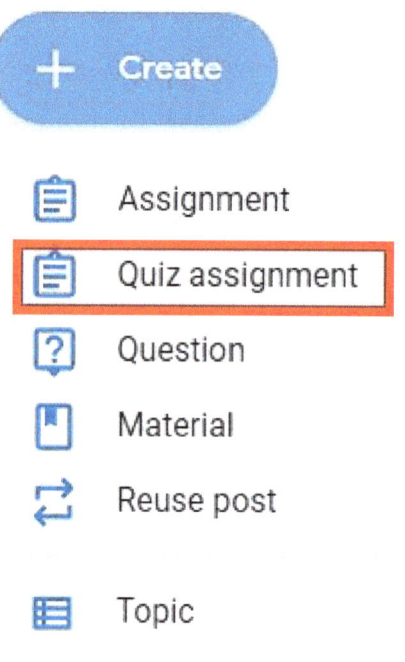

In the right part of the task screen you can decide:

- The course to assign it

- Students to assign it

- The maximum score

- The expiration date

- The argument (which can be easily created here).

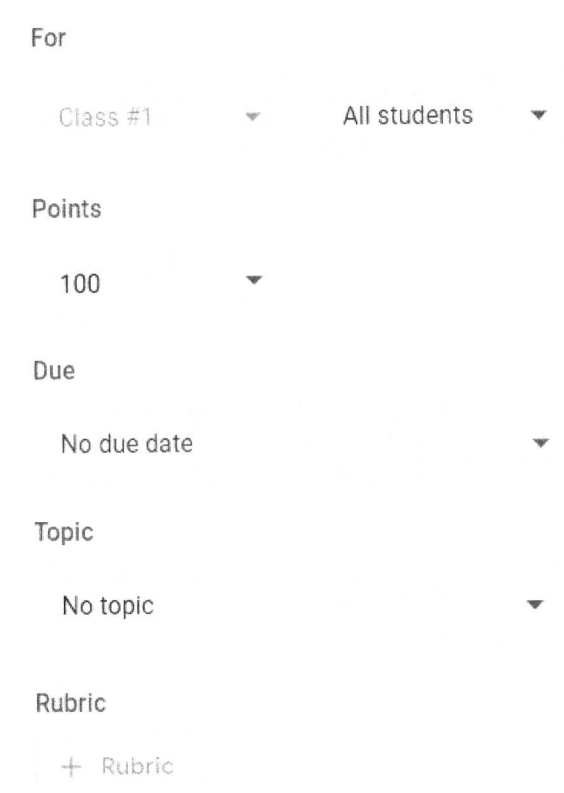

For

Class #1 ▾   All students ▾

Points

100 ▾

Due

No due date ▾

Topic

No topic ▾

Rubric

+ Rubric

In the left part, you can enter the title, task instructions and you can decide whether to set "Import votes"(to select if you do not want to have a quiz with anonymous answers). You can also add links, materials from drives, files, YouTube videos, or create documents, sheets, presentations, etc., by simply clicking on the quiz icon you can access its composition.

To define the questionnaire, you must activate the change by clicking on the pencil symbol at the bottom right. To define the questionnaire, we must select the different possibilities that Google offers us. We can choose whether to do multiple choice quiz, free answer questions, or checkboxes.

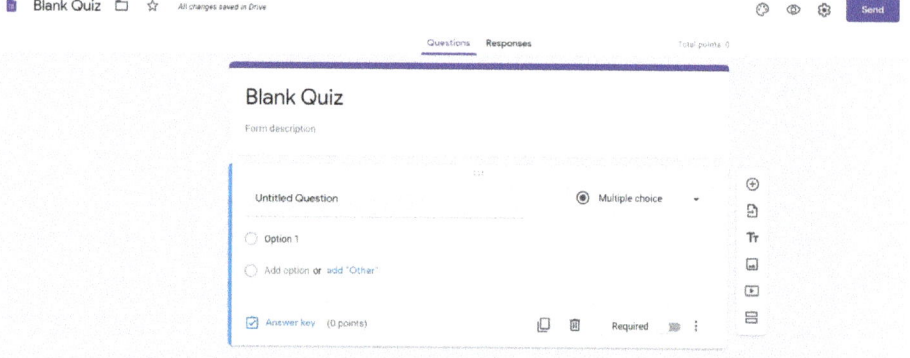

You can, of course, change the title and enter a description. If other materials have been added to the task, the "import votes" option seen earlier will be disabled, and therefore, to not having an anonymous questionnaire, you need to click on the settings and check the choice: "collect email addresses", remember that. Also, an important section is a vertical toolbar on the left with which you can: add a question, import questions, and add a title, image, video, or section.

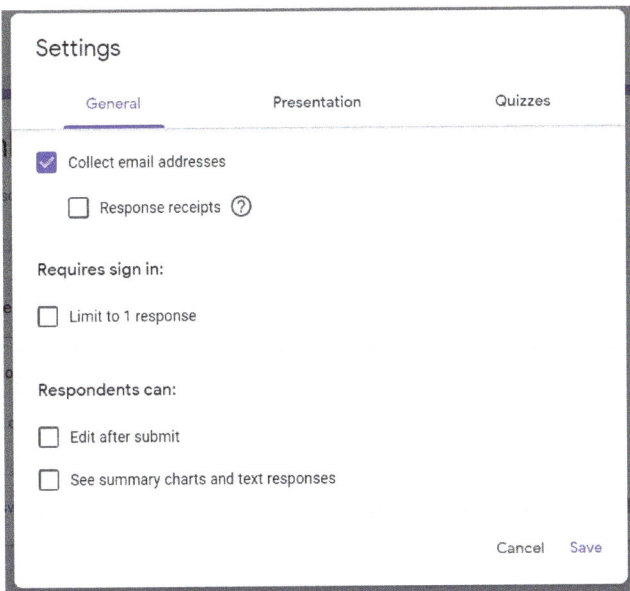

By clicking on the question you access its definition.
A multiple-answer question is proposed, but the typology can be changed with the menu on the right. For multiple-choice you enter the question with the desired response options. With the "show" menu at the right bottom (the three dots), you can enter a description of the application and have the answer options arranged randomly. For this type of question is then fundamental click on "Answer key."

After clicking on the "Answer key", you can select the correct answer, decide the score for this question and, if you want, add feedback. Of course, if we select multiple answers in the answer key, the exact choices can be more than one. Concluded those transactions, click on "Done."

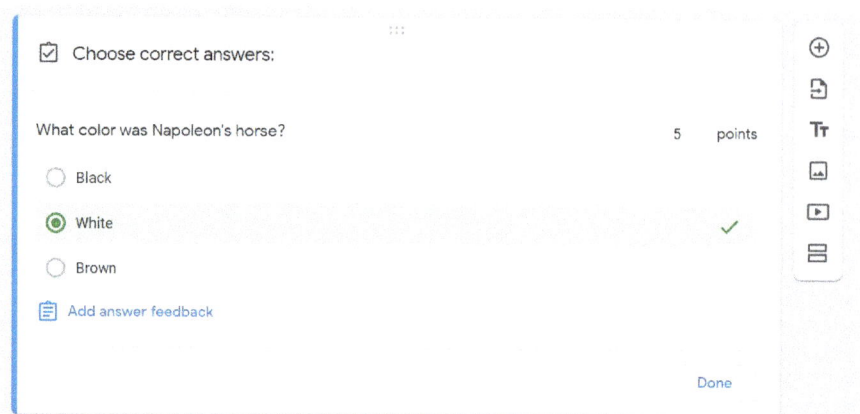

For checkboxes, you can enable the "Answer validation" from the "show" menu (three points), by entering the rule for response and an error message.

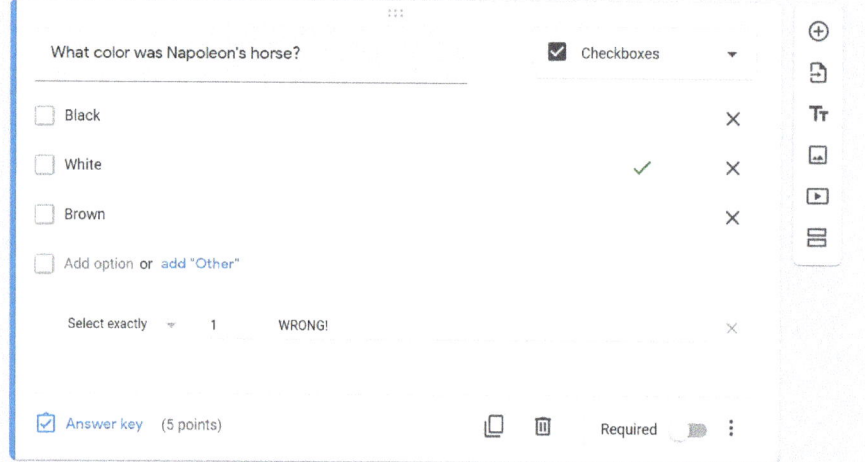

Also for open questions, you can activate the "Answer validation" from the "show" menu, entering the rule for the answer (a minimum or a maximum number of characters, or a pattern error message).

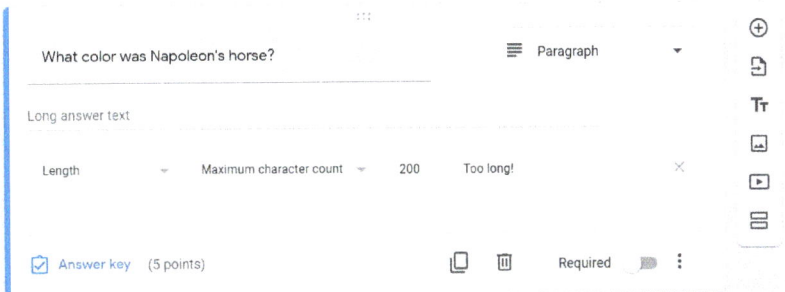

Obviously, in the "answer key" for open questions, you do not have to select no exact answer, and it is, therefore, sufficient to enter the score and possible feedback.

Final Retouches

Once you enter the questions you can customize the theme assigned to the questionnaire and preview. You can choose different colors to customize your tasks; Google Classroom offers a wide variety of customizations.

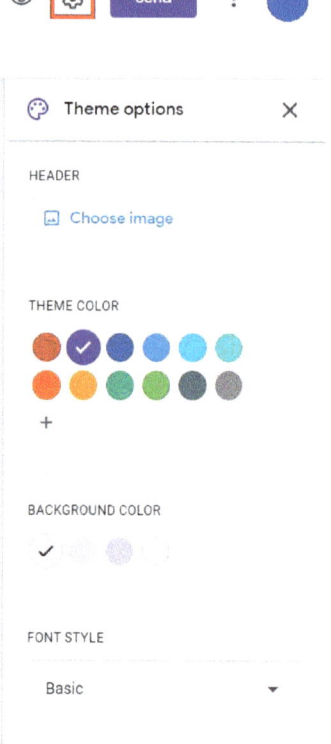

Once all operations have been completed, you can leave the definition of the questionnaire. The task appears in the list of "course works" in the form of a draft and it is possible to edit it or assign it by clicking the "Edit assignment."

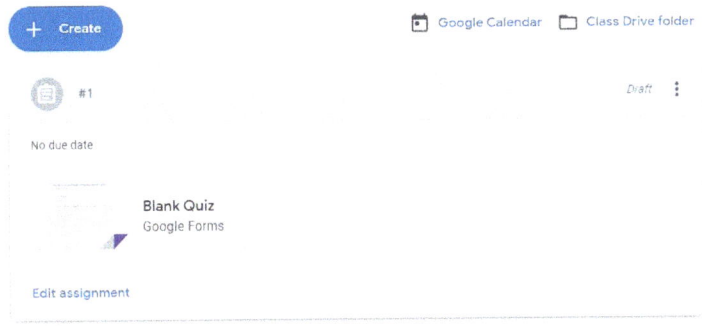

Clicking on "Assign" will assign the task, while activating the menu next to it is possible:

- To assign the task

- To schedule it for a certain date-time

- To save the task draft.

You can also use an evaluation grid by clicking on "Rubric." You can create, re-use, or import from Google Sheets, a grid; this is the screen for the creation.

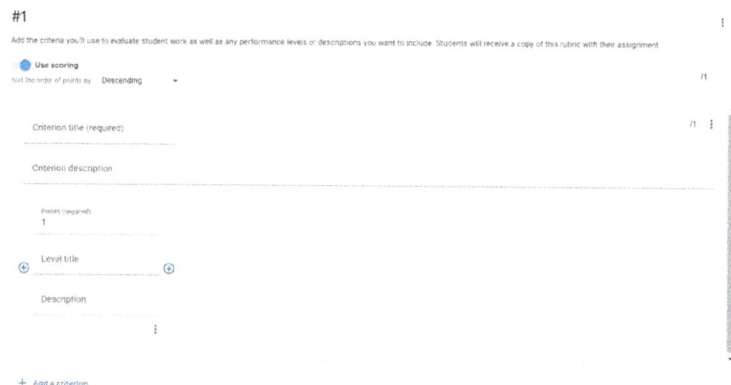

Once assigned the task, in the "Classwork" tab will appear on the publication date, the indication of how many students have had the allocation and how many of them have turned in.

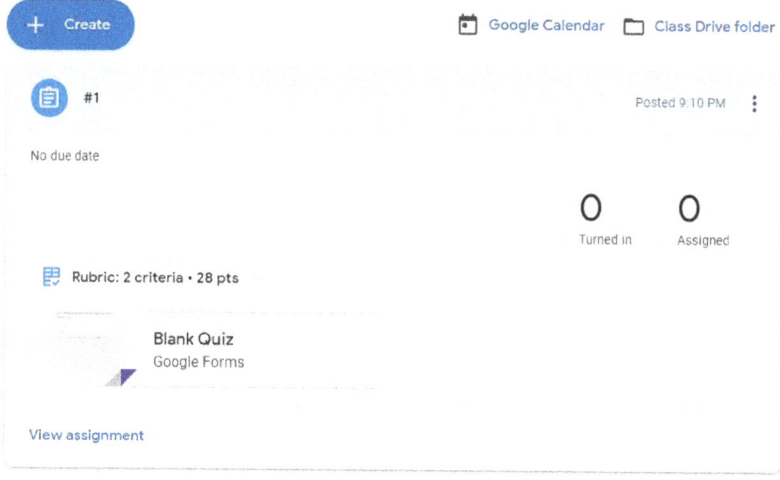

Cancellation of a course

To delete a course you must first archive it.

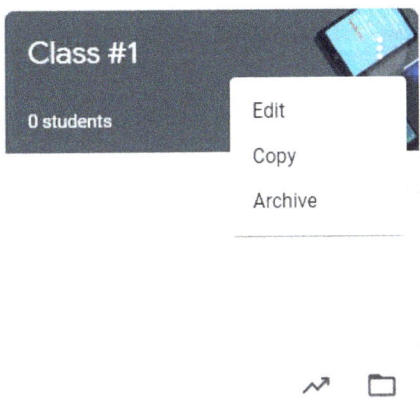

Later you can:

- Click on the classroom menu

- Access archived courses

- Delete the course you no longer need

≡ Archived classes

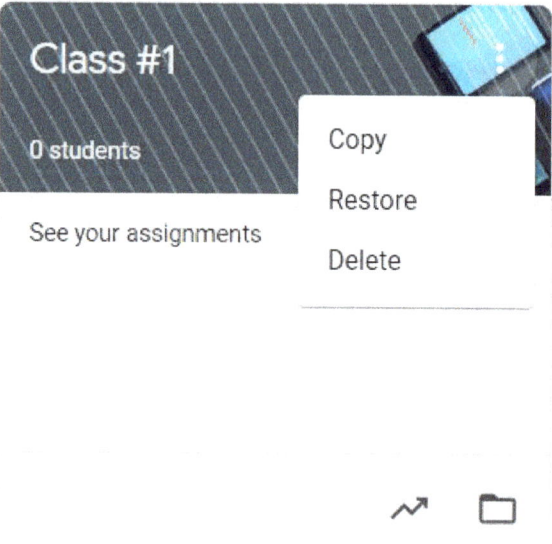

To exit Classroom (and the Google Suite platform):

- Click on the icon at the top left with the initial of the account

- Click "Exit."

Name and Basic Data of the Course

Customize the Class Color

Class themes and colors are something that you can integrate into the classroom. You can, with this, go to the settings, and then choose the default color or a theme for your class. This does help if you are working with multiple classes and want to make sure you provide the information to the correct class.

Name Your Class: To build a new class press the "+" sign.

Using the class name's rational naming structure and be consistent with all classes. Starter tip: make all your classes in backward order. For example, if you are teaching seven class periods, first build the 7th, then the 6th, then the 5th, and so on. This will enable you to view your lessons on the Google Classroom home screen in sequential order, which you will observe in this book later on.

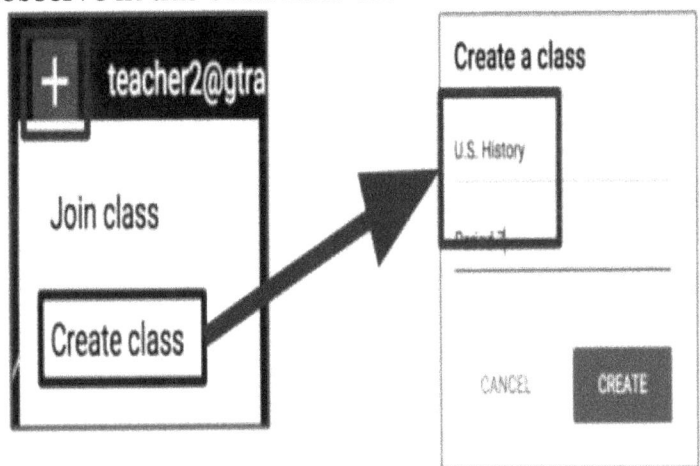

Take a tour: For students and teachers both, there is a useful tour developed into Google Classroom. Taking just a few minutes to get to know the application is worthy of your time.

Add Class Details

About tab: The About tab is where you insert your class details such as a summary of the course, curriculum, other learning materials, invite co-teachers, and much more.

Kasey Bell

Teacher

kasey@gtrainerdemo....

INVITE TEACHER

Invite Co-Tutor: In this section, you'll see the option on the left-hand side to consider inviting a co-teacher. To invite the co-teachers to your class, click on the INVITE TEACHER button provided.

Co-Teachers have the same authorizations as headteacher.

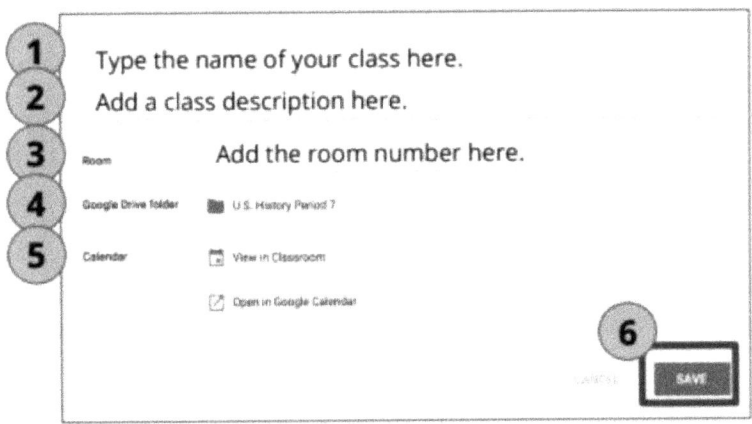

Add Details on Class

You may also add more class specifics and references to other information here like.

1. Your class title.
2. Class summary.
3. Classroom number.
4. Click to view the Class directory which was generated automatically in Google Drive.
5. Tap on a new tab to open your class schedule.
6. Click on Save to save your modifications.

Add Class Materials

About tab: This tab is a great hub for your teaching resources too.

Please remember to add files such as:

- Syllabus of your class.

- Rules and Regulations for your class.

- You can reuse the Google Doc, Google Slide, or Google Drive folders as well as many other documents or files that learners will require during the school year.

Add Materials

For adding files to your class, click on Add materials option.

Add materials...

Insert the label of your materials and additional information and documents from either your local storage or from Google Drive, YouTube videos, or links. Click the POST button to publish.

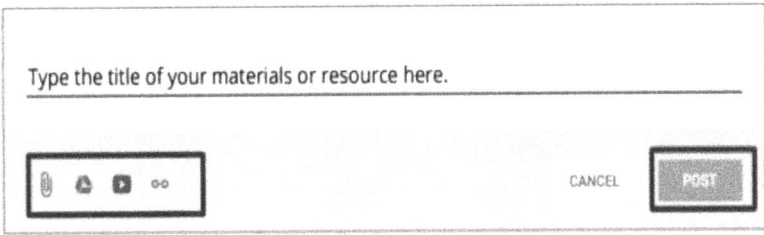

Useful Tip!

All the most essential materials, year-round, should be at the About tab.

If this segment starts growing too big, it would be difficult to locate some of the files or links that may be important.

Try sharing a Google Document, a Drive folder, or a website link for more substantial and structured resources in the About tab so learners can find out what they require.

No justifications!

It is time to add to the students, and it is not going to be much of a classroom. Remember that you will need to go through these steps for each of the classes that you set up, so be careful that you are getting the right students into the right classes.

The Classroom Stream

Like many other Google items, Classroom offers a Stream of sequential posts for publishing assignments, queries, and other events. Scrolling was very time-consuming, as more posts were added, according to educators.

Now, a new section on Classwork organizes content by modules and units, making it easier for teachers to plan their curriculum by semester and unit, as well as allowing students to find assignments more quickly.

The stream is the arrival area for Classroom that permits quick access to every one of your records and where everything going on in your group can be shown—it's what your students will initially observe when they sign into Classroom. That is the reason it's an extraordinary device for posting new assignments. It's likewise a decent method to get the class to associate.

You can transfer specific inquiries to be replied by students in Stream. You can likewise put assets, for example, reports, pictures, and even recordings that permit students to get a more extravagant wellspring of data identified with your course.

In the Settings segment (the machine gear-piece at the upper right), you can pick how students connect with "Students can post and remark," "Students can just remark," or "No one but instructors can post and remark" alternatives.

You can likewise choose what appears on the Stream with choices of "Show connections and subtleties," "Show dense notices," and "Conceal notices." You can likewise permit this to show erased things by utilizing the switch button alternative.

By giving the alternative to students to remark, it's feasible for educators to get criticism on how assignments and assets are being acquired—taking into consideration future work to be engaged such that will better connect with the students.

The Stream demonstrations like a course of events, which implies the two students and instructors, can look back through the work that has been set, permitting everybody to discover more seasoned assets. This is an invite expansion with regards to overhauling time, as students can look for some kind of employment all the more autonomously and let loose the educator to concentrate on different assignments. We're not saying no students will ever email an educator regardless of the work being in that spot, yet it should make it somewhat more transparent for everybody.

The Stream is now being updated for tasks with a streamlined perspective that highlights queries and more conversations. This makes the Stream a greater hub of conversation, where instructors can inform their students of upcoming deadlines, posting announcements, and more. Students may comment on the messages, establishing a Classroom virtual forum.

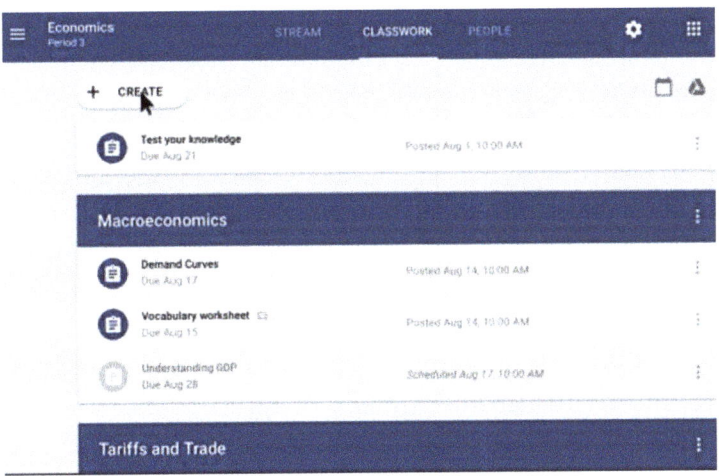

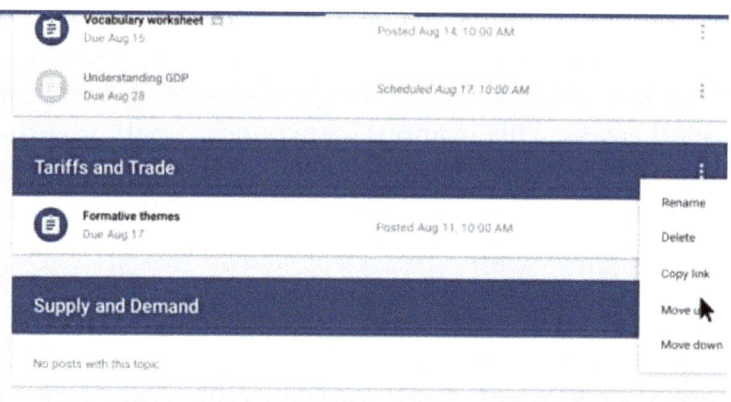

Certain features coming this fall include an updated understanding of the community that helps students, co-teachers, and guardians to be properly managed. The page helps teachers to introduce and delete accounts, change details regarding the guardians, and submit emails. Settings were centralized, as well. They can modify the overview of the class, alter the code of the school, adjust the summaries of the guardian and the position of the class, and monitor how the student's comment and post on the Stream.

The other major change that will arrive later this year is the opportunity to lock the screen of a Chromebook while a quiz is running. Until all answers are submitted, students will not be able to switch to other apps or tabs. The new cheating prevention is only accessible on controlled Chromebooks, which will begin this fall with the option to build a Quiz right from the Classroom, rather than the Google Forms program.

The Stream contains all chronologically ordered posts, from the newest to the oldest.

Add To the Stream

To add anything to the stream, click the + in the bottom right corner. You can create an assignment, announcement, question, or reuse a post.

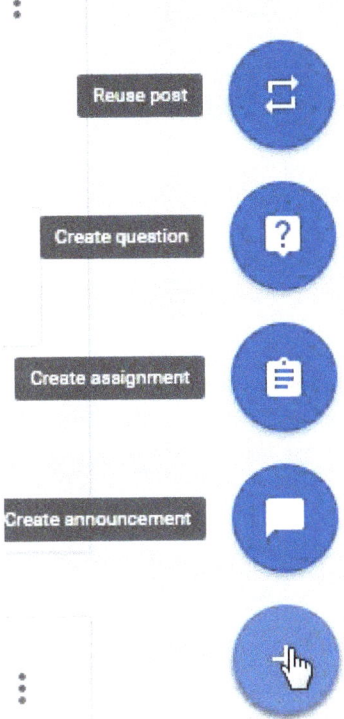

Stream Filter

By attaching the topics to your posts, the Stream can be filtered. Themes allow you and the students to filter the Stream to see related posts to that topic. You may create topics for every chapter or assignment types, for example.

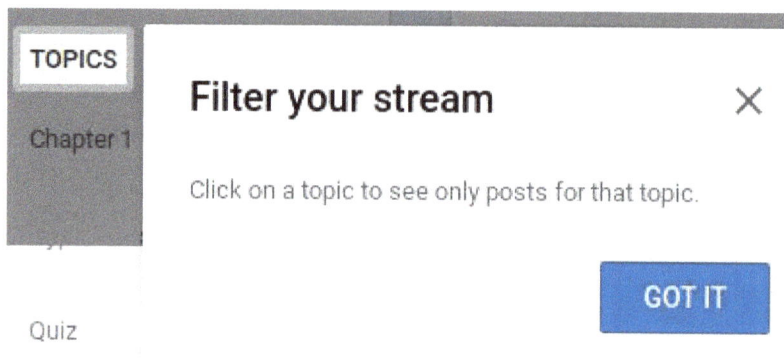

Post Schedule

You can make and upload a post and it goes to the Stream immediately, or you can plan the post to appear at a certain time and date.

Reuse Post

You may recreate a post from either existing or archived classroom. Don't delete Stream posts even if you think Stream gets too long. If you remove a post, you do not have it to respond to later, so you will not be able to reuse it.

Return Post to Top

You can move any post up the Stream to the top to draw attention to it.

Share the Course Code to Give Students Access

There are different ways of how your students will get access to your class. You may give them an invitation email to enroll students in the class or exchange the code of the class.

- Invite email—Students can press "Join" in the email or the class card after you have sent the invite.

- Class code—Students insert it in Google Classroom after you exchange the code for joining the class.

If students have difficulty with a class code, you also can reset it, or give them an invitation via email.

Note: Students may unenroll from classes. If they do so, their grades would be deleted.

Send an Invitation via Email

You may submit an invitation email to the specific students or a student group.

Note: You will use the e-mail alias to invite a student group to Google. You do not need to be a group member or owner, but you need to be able to see group members and the email addresses. If this information is not viewable, ask the administrator to adjust the permissions.

1. Visit classroom.google.com

2. Click the class you'd like to invite to.

3. Tap on People at the top and then click Invite students.

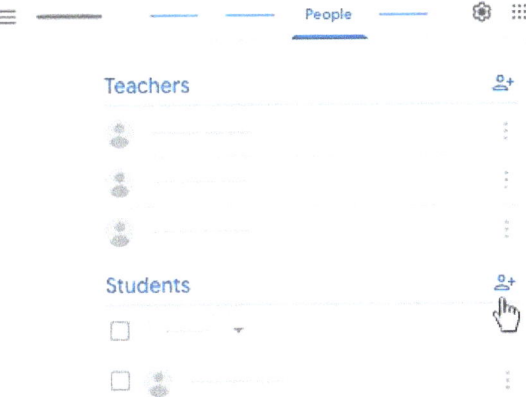

4. Enter a student or group's email address.

When you enter code, a list of autocomplete results will appear under the Search Results.

5. (Optional) Click on a student or group under the search results.

6. (Optional) Follow steps four & five, to invite more groups or students.

7. Click the Invite button.

Note: If you are having trouble adding the email address, this may be outside the domain of the school.

When you have sent out the invitation:

- It updates the class list to show the names of the students allowed to attend.

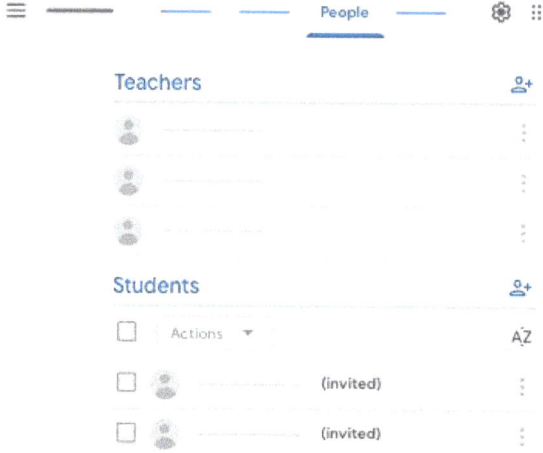

- Press "Join" in the email or on the class card.

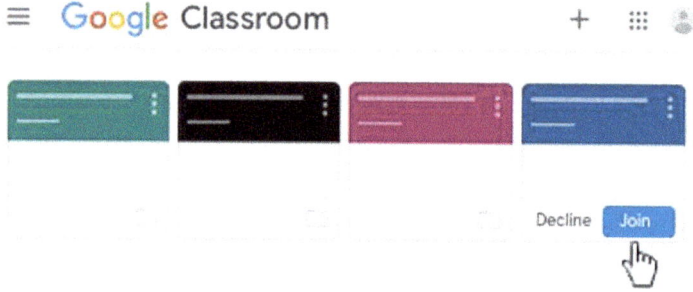

Share the Code for Class

Each class that you create has a code for the class. You share the code with the students to allow them to enter your class. Paste it in a message to share the code, display it with the projector or write it on some board.

1. Visit classroom.google.com

2. Tap the Settings.

3. (Optional) for copying the code so that you can insert it in a post, click on the Down arrow then click Copy under General, next to the address.

4. (Optional) Click Down arrow next to the code to display the code with the projector, under General.

Click the Full screen to get a full view of the code.

Tip: The code can be viewed easily from the Stream page. Click the Fullscreen next to the code under the class name. Click the Fullscreen again to get a larger view.

5. Offer such instructions to the students:

- Visit classroom.google.com

- Click Add Join class on the Classes page.

- Type the code in and press Join.

Invite Pupils From Outside the Classroom

You may welcome students from beyond the domain of the school or organization the same way you accept students from your domain, but the setting needs to be switched on by your IT administrator.

When you are an administrator, and then go to Configure class settings to switch on or off the setting.

Reset, Turn Off, or Copy the Class Code

When you make a lesson, Classroom generates a code for the class automatically. You should restart it if the students have problems with the class code. You can disable the code if you don't want to see new students entering your class. It can be enabled at any time.

Note: Classroom generates a new code for that class when you reset or activate a class code. The preceding code no longer works.

1. Visit classroom.google.com
2. Click the Settings.
3. Under General, click the Down arrow next to the class code, and select the option:

- Click on Display to view the code.

- Click on Copy to copy code.

- Click Reset, to reset the code.

- Press Enable or Disable to turn code on or off.

4. Click the Save button.

The "Class Work" Section How It Works

Classwork is the segment in Google Classroom wherein you can make assignments straightforwardly. Sounds like the Stream area, isn't that so? The distinction here is that you can set focus esteem or select ungraded, and dole out a due date for the work and a point for the task.

The extremely extraordinary part about this area is that you can post the task quickly or plan it to go out sometime in the not too distant future. Or, on the other hand, just spare it as a draft to finish it sometime in the not too distant future. This implies you can arrange work early, which permits you to be progressively adaptable with your time — perfect in case you're stuck at home.

An educator needs to explore Classwork. At that point, select the in addition to the symbol where it says "Make" and pick what will be set, from Assignment, Quiz task, Question, Material, or Reuse post. For this situation, select Assignment.

You, at that point, have the Assignment center point to enter a title and what the errand is. Put addresses in that spot into the container, permitting students to transfer in a Doc, or include your own Doc with the inquiries in it. Or on the other hand, add a link to a site with the surveys, or even insert a YouTube video in that spot.

When it's good to go, you can include the students who the task is for starting from the drop menu on the right, choosing people, or all students. Set the number of focuses the assignment is worth and include a due date, likewise both on the right. When posted, this will show up in the Stream immediately.

Additionally, in the Classroom, you can make a Quiz that consequently provides a Google structure format that you can use for your students to fill in. This allows you to spread out inquiries with a selection of answer options that students can choose from. This is a simple method of having the answer in place within Classroom, rather than using extra Docs. It also allows you to get a total consistent evaluation. Just select Blank Quiz, include the inquiries and answer choices (in case you're utilizing different decision), or look over different alternatives, including short answer; section; checkboxes; dropdown; record transfer; linear scale; various decision network; and checkbox lattice.

Questions are another creative alternative that permits you to suggest a conversation starter and have students to react and can be open for others to see. Hit the Create button, pick Question. At that point, including the inquiry and how it's replied. However, there are many equivalents to the above questionnaire with one query rather than different ones.

Materials are an incredible alternative to use to impart assets to students. This makes for simple access to records on your gadget or puts away in Google Drive. Hit Create, and you can pick Material, include a title, portrayal, and afterward utilize the paper cut symbol to add any connections you need from Google Drive, Link, File, or YouTube.

Reuse post is, as the name proposes, an approach to take a past post and use it again—this can mean sparing a duplicate to make a comparable piece with another subject, say, without losing the first.

There are likewise Google Calendar and Drive tabs that permit you to see all due task dates and any assets that you've joined us you've come.

Divide the Contents into Topics / Modules

How to organize?

The topics are the core of the current Classwork website. Topics let you divide tasks into groups.

New topics are applied to the bottom of your classwork page by default, so this involves shifting the topics up or down to match the needs. It seems to be the teachers' greatest problems using the latest Classwork website. Rearranging the topics and activities should be one of all such choices because you do not want students to drop down to the end of the list.

You do have the option of attaching materials to your Classwork list. It is usually the tools that you connect with students who are not automatically related to one task. There are items that students ought to regularly consult, such as the class syllabus, website link for instructors, and guidelines for courses, etc.

The following are a few successful methods of arranging tasks at Google Classroom:

Organize According to the Learning Target:

Assign letters and numbers to each learning goal and include them in the subject and task. Even if you don't have standard-based scoring, this is a perfect opportunity to help the students align the task with the task target.

Organize According to Modules or Chapters:

To arrange by study class, teachers develop themes for each class and bring them all together assignments within the theme for that specific organization. With the redesign, this approach was the original purpose of Google.

Organize According to the Type of Assignment:

Teachers arrange tasks in this system by job form, such as "Daily Work," "Projects," "Tests," etc. Know the company for you and students alike. Can students comprehend the styles of work? With the age level you are teaching, try to make things student-friendly.

Organize According to Subject Areas:

For the elementary teachers, who teach several topics, grouping by topic area appears to be the most appropriate.

Organize According to Weekly Basis:

For each week, several teachers develop a new subject, either calling them "Week # 1," or calling it like "Week of Sept. 10-14." If you think this can work then you can certainly do it.

Organizing by Adding Topic of the Day:

Normally, this approach is paired with any of the strategies above. The concept is to establish a specific subject called "Topic of the Day or Today," where anything you need your students to concentrate on the day is transferred manually. Then push the "Today" issue to the top of the page and it's for the students at the front and middle.

Organizing by Adding Topic of the day:

One aspect that appears to make some teachers nuts is that the tasks are always revealed in the "Stream." If it is a pain in your hand, this function may be deactivated to prevent misunderstanding.

Go to your class to deactivate this feature and click on the Settings icon which is near top-right.

Then scroll down to "General."

Click the drop-down button next to "Classwork on the Web," and click "Hide Notifications."

Make a Topic for Past Assignments:

Create a topic which is called "Past Assignments" or something which makes sense to understand both your students and you. Drag the subject down to your Classwork tab. When an assignment has been finished, or you no longer approve work, transfer it to this topic.

Creating a Task to Be Assigned To Students

Assignments are a useful tool on Google Classroom for delivering, tracking, and also grading student submissions. Even non-electronic submissions can also be tracked using the Assignments tool.

Add an assignment

Creating an Assignment:

- Open classroom.google.com.

- At the top, click on Class and open Classwork.

- Also, click on Create and click on Assignment.

- Input the title and necessary instructions.

Posting assignment:

A. To one or more classes:
- Just below for, click the drawdown on Class.

- Choose the class you want to include.

B. To individual students:
- Select a class and click the drawdown on All Student

- Uncheck All Student

- Then select the particular student(s).

Inputting grade category:

- Click the drawdown on Grade Category

- Select Category

- Edit the following (Optional):

- Click Grades to edit the grades page

- Click Instructions to compose the Assignment

- Click Classwork to create a homework, quiz, and test.

Change the point value:

- Click the drawdown below Points

- Create a new point value or click Ungraded.

Edit due date or time:

- Click on the drawdown below Due

- Click on the dropdown on No due date

- Fix date on the Calendar

- Create due Time by clicking Time, input a time adding AM or PM.

—

Add a topic:

- Click on the drawdown below Topic

- Click on Create topic and input the topic name

- Click on an existing Topic to select it.

Insert Attachments

File:

- Click on Attach

- Search for the file and select it

- Click Upload.

Drive:

- Click on Drive

- Search for the item and click it

- Click Add.

YouTube:

- Click on YouTube

- Type in the keyword on the search bar and click search

- Select the video

- Click Add.

For video link by URL:

- Click on YouTube and select the URL

- Input the URL and Add.

Link:

- Click on Link

- Select the URL

- Click on Add link.

You can delete an attachment:

- Click removes or the cross sign beside it.

You can also determine the number of students that interacts with the attachment:

- Click on the drawdown besides the Attachment

- Select the required option:

- Students can view file – This implies that students are allowed to read the data but cannot edit it.

- Students can edit the file – This means students can write and share the same data.

- Make a personal copy of each student – This means students can have their transcript with their name on the file and can still have access to it even when turned in until the teacher return it to them.

Note: If you encounter an issue like, no permission to attach a file, click on copy. This will make Classroom make a copy, which is attached to the Assignment and saved to the class Drive folder.

Add a rubric:

You must have titled the Assignment before you create a rubric.
- Click the Add sign beside Rubric

- Click on Create rubric

- Turn off scoring by clicking the switch to off, besides the Use scoring

- Using scoring is optional, click Ascending or Descending beside the Sort the order of points.

Note: using scoring, gives you the room to add performance level in any, with the levels arranged by point value automatically.

- You can input Criterion like Teamwork, Grammar, or Citations. Click the criterion title

- Add Criterion description (Optional). Click the Criterion description and input the description.

Note: You can add multiple performance level and Criterion

- Input points by entering the number of points allotted.

Note: The total rubric score auto-updates as points are added

- Add the Level title, input titles to distinguish performance level, e.g., Full Mastery, Excellent, Level A

- Add Descriptions, input expectations for each performance level

- Rearrange Criterion by clicking More and select Up or Down

- Click Save on the right corner to save Rubric.

Reuse Rubric:

- Click on the Add sign beside Rubric

- Click Reuse Rubric

- Enter Select Rubric and click on the title. You can select a rubric from a different class by entering the class name OR by clicking the drawdown and select the Class.

- View or edit rubric, click on preview, click on Select and Edit, save changes when done. Go back and click Select to view.

View rubric assignments:

- Click on Rubric

- Click the arrow down icon for Expand criteria

- Click the arrow up icon for Collapse criteria.

The grading rubric can be done from the Student work page or the grading tool.

Sharing a Rubric:

This is possible through export. The teacher creates the Rubric exports, and these are saved to a class Drive called Rubric Exports. This folder can be shared with other teachers and imported into their Assignment.

The imported Rubric can be edited by the teacher in their Assignment, and this editing should not be carried out in the Rubric exports folder.

Export:

- Click on Rubric

- Click More on the top-right corner and enter Export to Sheets

- Return to Classwork page by clicking close (cross sign) at the top-left corner

- At the top of the Classwork page, click on Drive folder and enter My Drive

- Select an option, to share one rubric, right-click the Rubric. To share a rubric folder, right-click on the folder.

- After right-clicking, click on Share and input the e-mail you are sharing to.

- Then click Send.

Import:

- Click on the Add sign beside Rubric and enter Import from Sheets

- Click on the particular Rubric you want and click on Add

- Edit the Rubric (Optional)

- Click on Save.

Editing Rubric Assignment:

- Click on the Rubric

- Click on More at the top-right corner and enter Edit

- Click Save after making changes.

Deleting Rubric Assignment:

- Click on the Rubric

- Click on More at the top-right corner and enter Delete

- Click Delete to confirm.

Posting, Scheduling or Saving Draft Assignment

Post:

- Open Classwork and click on Assignment

- Click on the drawdown beside Assign, on the top-right corner

- Click on Assign to post the Assignment.

Schedule:

- Click on the drawdown beside Assign, on the top-right corner

- Enter Schedule

- Input the and date you want the Assignment posted

- Click Schedule.

Save:

- Click on the drawdown beside Assign, on the top-right corner

- Enter Save Draft

- Editing Assignment:

- Open Classwork

- Click on More (three-dot) close to Assignment and enter Edit

- Input the changes and save for posted or schedule assignment, while Go to Save draft, to save the draft assignment.

Adding Comments to Assignment:

- Open Classwork

- Click Assignment and Enter View Assignment

- Click on Instructions at the top

- Click on Add Class Comment

- Input your comment and Post.

To Reuse Announcement and Assignment

Announcement:
- Open the Class

- Select Stream

- Slide into the Share something with your class box and click on a square clockwise up and down arrow or Reuse post.

Assignment:

- Open Classwork and click on Create

- Click on a square clockwise up and down arrow or Reuse post

- Select the Class and Post you want to reuse

- Then click on Reuse.

Delete an Assignment:

- Open Classwork

- Click on More (three-dot) close to Assignment

- Click on Delete and confirm the Delete.

Creating a Quiz Assignment

- Open Classwork and click on Create

- Click Quiz Assignment

- Input the title and instructions

- You can switch on Locked mode on Chromebooks to ensure student can't view other pages when taking the quiz

- You can switch on Grade Importing to import grades.

Response and Return of Grades

Response:

- Open Classwork

- Click on Quiz Assignment and free Quiz Attachment

- Click on Edit and input Response.

Return:

- Open Classwork

- Click on Quiz Assignment

- Pick the student and click on Return

- Confirm Return.

Overview of the Types of Content We Can Upload: Videos, Docs, Forms, Drawings

Google Drive

This app is a kind of free Cloud Storage for saving all of your files to enable easy accessibility (up to a storage capacity of 15 gigabytes).

I know that tons of you have heard the term "Google Drive," but did you furthermore may see that it comes with a vast 15 gigabytes of free storage?

It's one of the simplest tools available in recent times. Through Google Drive, as you've got good internet access, you'll easily save your files via this massive 15-Gb cloud storage.

Some of you would possibly be wondering, "What quite files are often stored on Google Drive?" Well, various file formats starting from Microsoft Word files, audio files, videos, PDF files, images, and then many other sorts of file formats are often stored on the drive.

The vital importance of using Google Drive is for straightforward accessibility, effective communication, and secure distribution of files and you'll distribute data to your students so that they will utilize them how they like.

Google Drive is the best organizer when it involves natural and secure file accessibility. For teachers, students, colleagues, and everybody else.

Classic Tips to assist Teachers: Google Drive is cloud-based storage, but that does not mean that it can only be accessed via the web. You'll also configure your Google Drive for offline use, and for you to be ready to access your files when an online connection isn't available.

Google Docs

This program is the most straightforward Microsoft Word alternative and is free and accessible online.

Many people confuse Google Docs for Google Drive, others think they're similar, but during a real sense, they're considerably different.

The difference is that Google Drive is employed for storing your files online. In contrast, Google Docs is beneficial for creating documents that are almost like Microsoft Word documents. Despite the similarities between both Google Docs and Microsoft documents, Google Docs differs because it's free and cloud established.

Google Classroom has made paperless writing an opportunity; the times of "the dog ate my homework" are long within the past. Students can submit their assignments, term papers, projects, tests, book reviews and tons more via Google Docs.

It makes the method of creating and editing documents and notes from the comfort of your computer quicker, as you don't need to hover over an outsized pile of paperwork for ages trying to sort them out accordingly.

Google Docs also can be utilized during a team-up writing task, where a specific team of scholars is expected to figure together. The in-built chat components are used for live teamwork task execution.

Google Forms

Google Forms is usually used for creating tests and other sorts of evaluations for college kids. The interesting thing about Google Forms is that these tests are used for the individual grading of scholars.

Google Forms, which may be a review tool, is indeed a superb resource. It's used for data collection, tabulating results, and an entire lot of other options.

As an educator, if you propose raising the extent of engagement for students' forms in your class, I suggest you try the Google Forms created by students.

Whether you wish it or not, every student in a method or another derives motivation from their peers. They're going to welcome the event of collecting peer data via a well-liked tool that's one hundred pc liberal to utilize.

There are many reasons why an educator should choose Google Forms. A number of these reasons are as follows:

- It's free to use.

- A section of the Google core suite is often easily formed and put in situ on Google Drive.

- Email messaging is straightforward.

- Google Classroom makes submission easy.

- When making use of Google-related sites, these forms are often embedded with ease across multiple sites.

This way is not a full list. As a matter of fact, the list is endless. Just touch navigation on your computer and any information is often made available on a Google sheet. Students can utilize this sheet to research and tabulate the info collected. This tabulated data is often pasted in similar sites, slides, docs, et al. As you'll see, you've got been ready to put together a lesson for the scholar, which has clothed to be interactive, captivating, and intriguing. During the utilization of the Google form provided by students. It's lovely, isn't it?

To create a private grading evaluation for college kids, you start by opening a replacement form in Google Drive. (Go to New > attend More > then Google Forms). Once you're done creating your test, you create a solution key.

Install the add-on Flubaroo to urge it functional. (Find you thanks to Add-on > get Add-on).

Tests are often given to specific students by using Flubaroo. Once the feedback has been received, you'll then send it to the scholars and their parents. Flubaroo is a crucial add-on, and it's free.

YouTube

Top-quality videos on every topic you'll imagine.

YouTube is the second hottest program in the world. One fascinating aspect of this video site is that you simply can find anything. YouTube is free, and you'll cash in by getting educational materials on an outsized sort of topics which will help when teaching your students.

All educational information is at your disposal, from Dalton's atomic theory to tons of documentary and history videos. Knowledge is endless on YouTube, which is why teachers must cash it in to make the simplest lesson plans for his or her students.

Google Drawings

Google Drawings is extremely much underrated by many of us. Google Drawings is supposed for any student, regardless of their educational levels.

Google Drawings enables users to form engaging and customized images and styles with ease. So, you'll utilize this tool to quickly support its simplicity.

Have you ever taught students, and at the top of the category, they spend time creating a manipulative, just for time to run out, and that they find yourself losing half it? Pretty often on behalf of me.

Drawings can overcome this problem. One of Google Drawings' primary applications is to use it for manipulatives — and if you're employed in high school or college and think that manipulatives are solely for grade school pupils, please join me.

You won't be ready to resist once you start exploring Google Drawings. You'll be using it to categorize polygons, objects, sentences, and endless lists. In Drawings, you'll see yourselves creating water cycle graphs, physical body structures, atoms, and you'll perhaps construct a variety table. There are infinite possibilities.

Just attempt to consider something Google Drawings can improve in your classroom. Then generate ideas, develop the lesson in your class, and execute it. I hope you and your students are getting to be focused on this application quickly.

Slides

Google Slides one of the foremost commonly used tools for presentations. It's an easy way of creating an introduction to an audience with reduced engagement.

One of the advantages of using Google Slides is that there are easy ways to get engagement and interaction together with your students.

Google Classroom Modules. To Ask Questions and Evaluate Students' Preparation Directly Online

Making Google Classroom modules will not only stop there. You have to take into consideration how your students will be engaged with the online classroom. With this, students will able to prepare how they will study and be active in class. Here are some student approach ideas that you can apply:

Suitable content and resources

When an assignment is created, there is a textbox available for the description of the task. This is useful and there is also an option to attach extra content that can be used as supporting resources. This can be utilized to help with the post to ensure students have all they need to approach the task at hand.

A good example of this is when adding the task; it is good to split the task into chunks that can be approached easily. This can be a lesson by lesson tasks that is broken into smaller steps. This could be done via videos, notes, questions, and clear instructions.

Supporting material can also be attached to ensure that students get all they need to understand the task. Screen captures are another useful tool that can be used to help with trickier content that needs detailed explanations.

Streamlining the content

Some students may require additional help and may get lost in the general class wall. Therefore, for some students, it is advised that they used private comments to submit and get feedback in their private space, so they do not get lost in the stream. As another approach, the assignments can be placed back to the top of their stream.

Feedback, feedback, feedback

When sending out work on the platform, you must implement good and thorough feedback to allow students to understand. This is a good way to establish good understanding so both instructors and students can understand where the student is and what is required to get them up to speed. When students can get instant feedback, this proves to be useful to self-assess their work. This is a great skill to develop when you can review with guidance and resubmit your work based on the feedback and improve your quality of work.

Section students based on their performance

One great approach to online learning is to utilize sectioning based on the performance of the students. Students over time will show their performance and then after some time, trends may be observed, and these students can be organized into sections based on their performance. There may be some students that perform highly and may need little to no supervision while in the class. These students may need to just do their tasks and submit their work and they are fine.

For others, they may need additional help and there may be the need to set up more material to help them. This material may have to be formulated in a detailed way to help them understand and see where they are falling short. Also, it is good to have students at the same level interacting and sharing ideas on how they have dealt with the topics and gained a better understanding.

Using creative ideas

There are creative ideas that can be used for students to be engaged online when using the platform. Videos, handouts, lectures, and various ways can be used. Applying the knowledge and show applications in different ways can allow students to appreciate and see knowledge working to help with understanding.

Brainstorming Sessions

This is a great idea to allow the users of the platform to brainstorm and share ideas in a session. This could be via social media groups, a learning forum to facilitate virtual brainstorming. The way of doing this can be done via providing topics or challenges that allow users to think outside of the box and relate deeper with the educational content. It can also be used to set up students in a group to assign a leader and help them to develop relevant skills to lead and allow all users to participate.

Question and Answer Sessions

Sometimes questions done via the platform can seem mundane. However, it should be mentioned that the right question can lead the cognitive thinking to overcome any learning challenges. One question can lead into the next; this can be a good way to engage and lead thinking and create lively online discussions.

Turning the tables

This is a great way to assign a student a topic and then allow then to create eLearning materials that can be shared with their peers. This can also be developed into a collaborative move that allows them to deep dive into topics and learns for themselves. Along with issued guidelines, the instructor can be clear what is expected of the students and give them tools to use. This is a good way of allowing them to learn by doing, this provides an alternative to just reading content and allowing them to fully understand the content so they can develop concepts of their own.

The Types of Questions That Can Be Entered In the Google Classroom Forms

How to Create Dropdown Questions

Step 1

First, open an assignment in Google Classroom, and then click on the gear in the upper right-hand corner, copy all grades to Google Sheets, and column D.

Step 2

You'll notice there is an arrow. Click on it to sort names from A through Z by the first name. Some students may have the same name, so add last initials, where appropriate. When you are done with the list, copy the names by using Ctrl C.

Step 3

Go to your Google Form select drop-down menu, and paste. It is as simple as that. When your students come to this form, they can select their name and move on easily.

How to Create File Upload Questions

Step 1

Go to your Google Form, go to the right and click on the pull-down, you'll see a new option in the middle of these question types that says "file upload." It says all files will be uploaded to the form owners Google Drive to make sure they only share this form with the people you trust. You can then click on "Continue."

Step 2

It will add that question in there, and you'll see that the maximum file size is 10 megabytes. Still, you can go ahead and change it to something other than that, if you want to up to the 10-gigabyte level. Hence, it depends on what you would like to do in terms of file size, especially if it's an image or something of that nature.

How to Create Linear Scale Questions

Step 1

So all you have to do is go to the plus sign and add a new question, and change it from multiple choices to linear scale. So you are going to go ahead and paste the question that you have already typed, then fill in your options.

Step 2

Next, you have the option of changing the scale from one to five, or one to six, or one to 10. And then all you have to do now is make a record if you want to. So when the student comes there. They can select how they feel on a certain skill set that you're evaluating, and they can rate it on the scale depending on how you set the scale rating. Maybe you want to see how the student feels about a certain standard or skill that they're learning in your class, and maybe this is a way for them to seek some help or guidance on that scale.

How to Create Multiple Choice Grid Question

Step 1

Go ahead and create the post in Google Classroom but as a reminder, if you want to add a question to the streaming Google Classroom, click the plus sign and choose 'create question.' Type the multiple-choice question that you want to ask your students.

Step 2

Go ahead and type in the question, if the question is self-explanatory, then there's no need to put in any instructions. And type in the choices that you give them in the multiple-choice answers that I gave them, you do have an option to allow students to see a class summary after they answer the question. So you may want to turn this on, if you're doing a multiple-choice question for a check for understanding or a bell ringer or an exit ticket that way the students can see how each other answered the question, with a question you can also attach a file or a Google Drive document, a YouTube video or a link to a website for students to refer to if you want to.

Step 3

So, as we know in Google Classroom, things don't always look the same way on the teacher side as it does on the student side, so go over to a student account and see how this question looks on their side. Now we can see that the student has already gone ahead and done those questions. They've already turned it in, but you can see there, the question will be displayed as well as the answer choices with radio buttons that will allow the student to select the correct answer. They will also have a submit button so that when they're ready to submit their answer, they can do so. So that is how a question looks on the student side, go back to the teacher side, and look at one more thing on the teacher side.

Step 4

Now, if one student has answered the question, you will see an overview. There is a class overview of how they are answering that particular question. If you want to see how a particular student answered the question, you are going to click on the student's name, and then you are going to be able to see her answer to that question. So with the multiple-choice question, you get to see an entire class overview. Still, you can also see how each student answered by clicking the student name to bring up their question.

How to Create Checkbox Grid Question

Step 1

Checkbox grid is very similar to the multiple-choice grid, and the only differences are that you can have multiple things check for each one. So there you can set up your different rows and also can set up your different columns.

Step 2

So with a checkbox stream that would allow you to enter more than one thing for each row, unlike what you did with the multiple-choice grid, where you can only have one or the other. You don't have a lot of different options there. You could do a description just like before, and you could limit responses to one per column.

How to Make Quizzes

Originality Report

Another important feature that you can use when creating your assignment is the Originality report. The Originality report is like anti-plagiarism software. This means that it compares the work of the student to others on the web, and notifies you when it detects the similar content elsewhere.

The Originality report also notifies you when the copied content isn't properly cited by the student. But you should note that the Originality report is only available for G Suite for education accounts; Personal Google accounts cannot utilize this feature. G suite for education accounts can only run this tool for three assignments per class, for every 45 days. They can only run the tool after every 45 days, and for 3 assignments only.

The G suite for education account has to be upgraded to a G Suite Enterprise for Education account, to be able to run this tool an infinite number of times. The students in a class can also use this tool, and only before submitting an assignment. They can run this report for a maximum of 3 times. Students can only use this tool after it has been activated by the teacher.

For individuals with the G suite for education account, they can easily turn on the originality report. You just need to go to the classwork page and try to create an assignment as you normally would. The originality report icon will be seen among the list of options there. You can activate it by clicking on the box next to it.

The originality report works automatically for the teacher when it has been activated. Thus, whenever a student submits their assignments, the tool automatically runs through it. The teacher can view the results of the originality report by navigating to the classwork page, clicking on "View Assignment", selecting the student's work, and clicking the "flagged passages" icon.

The teacher should take note that this tool does not in any way have the ability to detect the severity of plagiarism, as normal plagiarism software would. That part of the job is to be undertaken by the teacher. This tool does not run reports comparing the work submitted by students of the class, it only does so against websites accessible by Google Books and Google Search.

Rubric

The rubric is another important tool to be taken note of when creating assignments. It is a tool used to create conditions for grading homework. The rubric is based on solid logic, thus even parents believe it is an impartial tool for grading their ward's assignments. It is used to give a rating to the assignments done by students, to enable them to know and understand their level of progress in that class.

The criteria used for creating a rubric are important because the rating it will give to a student will point out, or expose all his weak points in his understanding of things taught in that class.

How to Create Rubrics

To create a rubric, you must first navigate to the classwork page. You can then click "create assignment", which opens up the interface for creating assignments. The option to create rubrics is only available when creating assignments. Click on the "Rubric" icon. This brings up a drop-down menu on which you can click to create a rubric, reuse rubric, or import from sheets.

If you click on the "Reuse rubric" option, it brings up a set of already created rubrics for you to choose from. The used rubrics can either be reused immediately, or they can be further adjusted before they are used. Do note that adjustments to the rubric being reused in a new assignment, does not affect the original rubric.

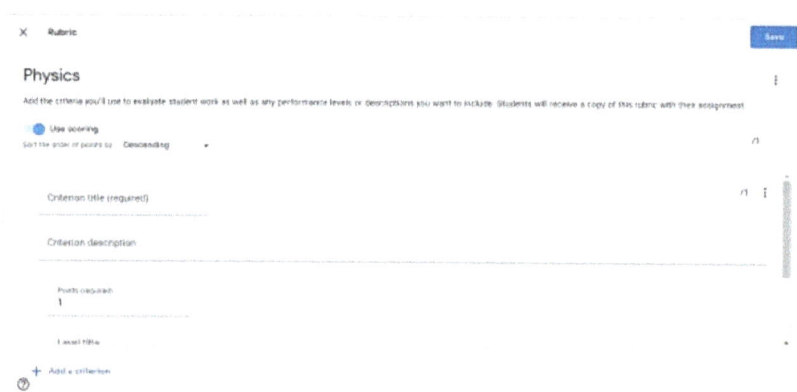

When you click on the "Create rubric" option, it takes you to a new page on which you can tailor the rubric you wish to create to your specifications. All the tools required to create a rubric will be available on the new page. You just need to click on the area you wish to edit, and input the details required.

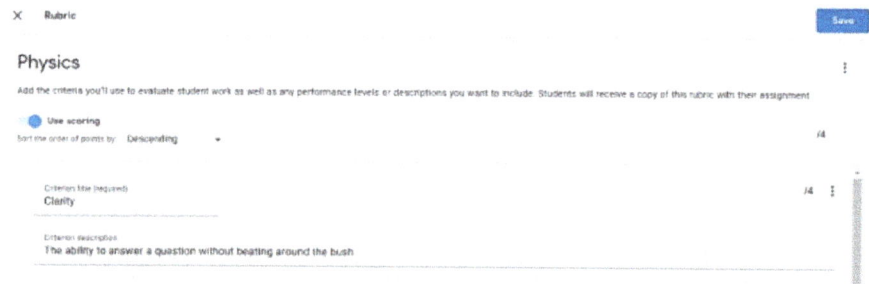

You can give a title to the criterion you wish to create, and also give a description to the criterion (this is optional). The criterion could be things such as punctuation, spellings, explanations, etc. It all depends on the subject in question, and the decision of the teacher. The criterion's description is just an explanation about the meaning of the criterion.

The next tool involves given grades to various performance levels. You can create as many performance levels as you need. The "Points" here indicates the total significance of that performance level, and it could be 7, 5, 8, etc. The "level title" denotes the name for that performance level, it could be titles such as poor, very poor, acceptable, good, etc. The description is just a short note showing the criteria for acquiring that performance level.

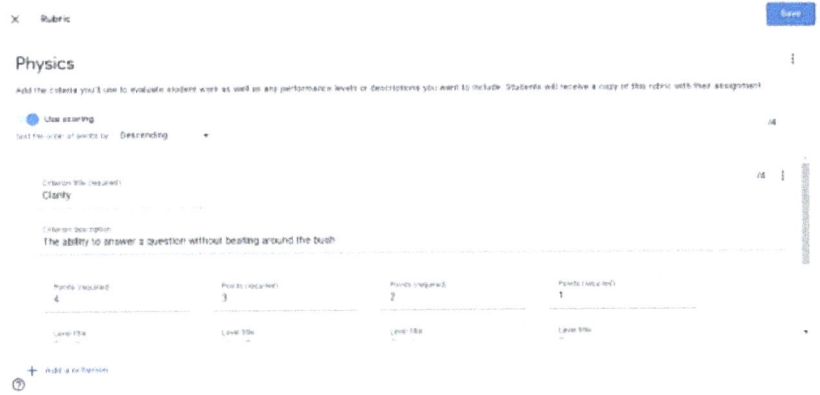

Several criteria can be created under one rubric. Thus, if you desire to add other criteria to the rubric, you can do so by clicking the "Add a criterion" button. This brings down another set of dialog boxes for you to create other criteria under the same rubric. After you are done creating the rubric, you can click the "save" button at the top of the page to save it.

When a rubric has been created and saved, its file format is the same as that of Google Sheets. Rubrics can also be imported from Google Sheets since their file formats are similar. But, the only Google Sheets that can be imported, and used in Google Classroom are those that were created in Google Classroom.

Creating Quizzes

A quiz is an assessment meant to test the level of understanding of an individual, and quizzes have always been a preferred method of most teachers in testing understanding. This feature is also available in the Google Classroom service. To create a quiz, you need to navigate to the classwork page and click on create.

The drop-down menu that appears will contain the icon "Quiz assignment", which you can click on to create a quiz.

This brings you to a page with similar features as that of the assignment. The process of creating the quiz is similar in all ways to that of assignment, except for one important feature. This means that you can specify the students whom you can send the quiz to, edit the points allocated to the quiz, edit the due date, add topics, add a rubric, and even add other attachments.

This means you can add YouTube videos, add website links, upload files from the computer's memory, etc. The title of the quiz is necessary, while the instructions are optional. But the one defining feature that separates assignments from quizzes is the fact that quizzes require you to use the Google Form service.

The last line on the page for creating a quiz shows an already prepared blank Google Form. Clicking on it takes you to a new page where you can customize the Google Form for your quiz. The main purpose of the Google Form is to aid the teacher in creating and structuring quizzes with multiple-choice questions.

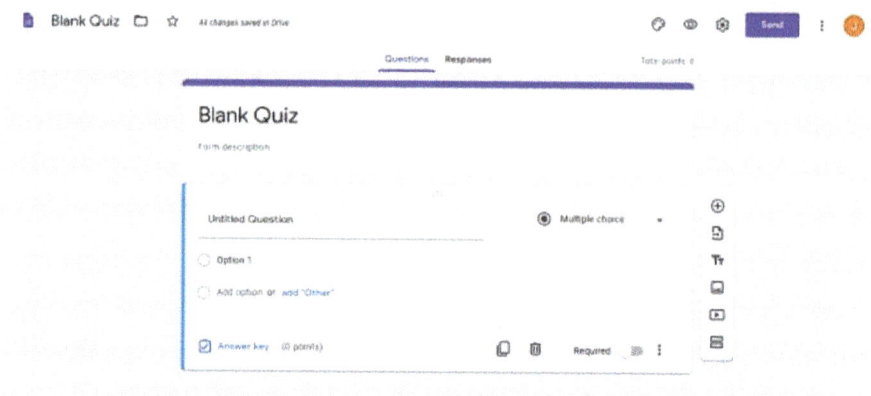

The "multiple choice" question format is the default layout of the Google Form, so no changes need to be made on this (But there are other formats available such as the "short answer" format). Click on "Untitled Question" in the first line of the form to type in the desired question. To create the options, just click on "option 1", and you would be able to enter the first option.

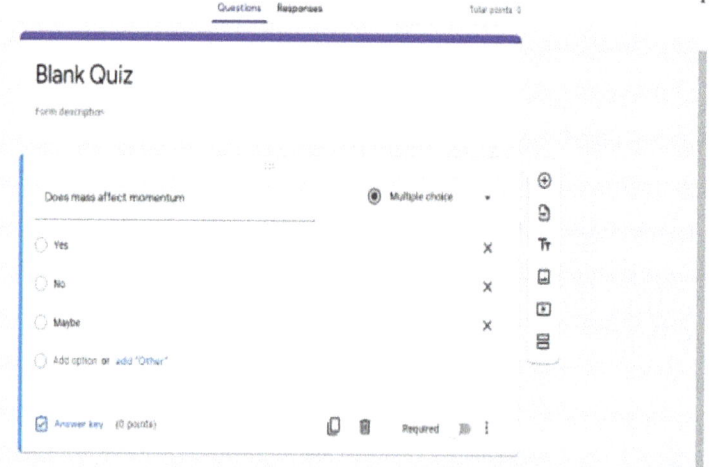

The other options can be entered in the "Add option" spaces below each filled out option. When you are done setting the options, there is a space beneath to set the points you wish to allocate to that question. After that, you can click on the "Answer key", which will show you a page where you can set the correct answers.

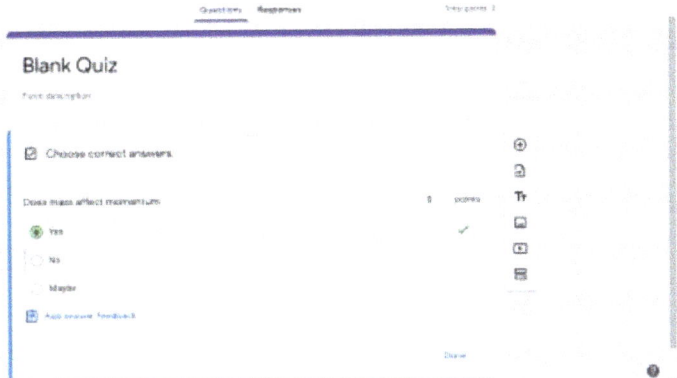

The new page enables you to select the correct answer for each question and allocate the points the correct answer carries. There is even an option to provide feedback for when the correct option or the incorrect options are chosen. After this, you can just click "Done" to finish the creation of that question.

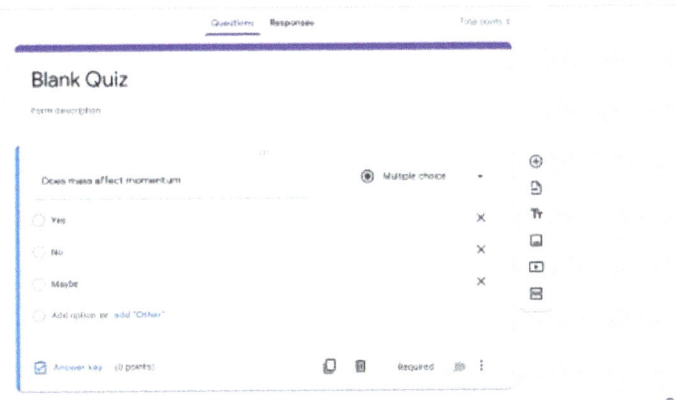

There is also a set of tools by the side of the quiz form, with options to add questions, import questions, add titles, add pictures, add videos, and even create a new section on the quiz. The quiz Google Form always automatically saves its contents, and as such when you are done, you can just close the form. After which you can go back to the classwork page, and assign the quiz, Schedule it for a specific time, or just save it as a draft.

To Create A Question

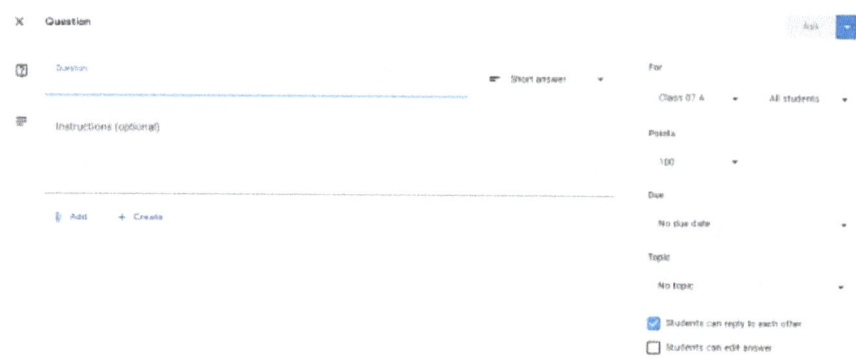

The steps taken to create a question are quite similar to those of the quiz, or the assignment. And this is because their layouts share lots of similarities. This means that you can also attach files to the question, specify the students you wish to post the question to, the class you wish to post the question to, the allotted points, the due date and time, the topic, the question's title, and its accompanying descriptions.

But there are a few things that make a question different, and this is because the question's layout has options for either choosing either a short answer format or a multiple-choice format.

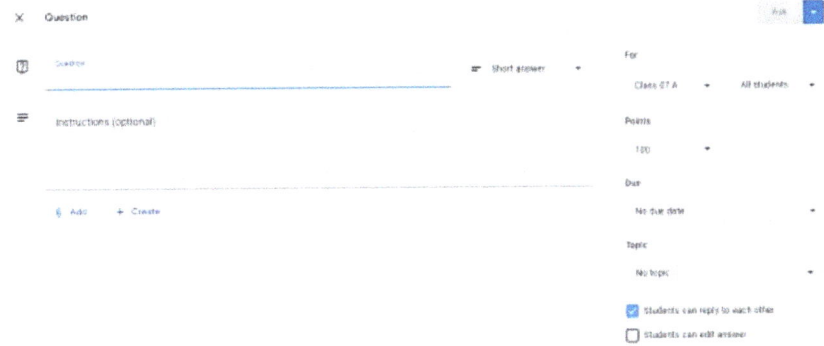

If you choose the "Short answer" format, this produces the feature of allowing the students to send messages to each other while they answer your questions, or if they can edit their answers after they have sent it to the teacher.

If you choose the "Multiple choice" format, this brings out an interface for you to enter the multiple-choice questions. It also produces a feature from which you can decide if the students will be able to see the class summary.

After making your decision on these features, you can then immediately ask the question, schedule it to be asked at a specific time, or just save it as a draft.

Add Multimedia Elements to the Tasks and Contents of the Course in General

Attaching a YouTube Video

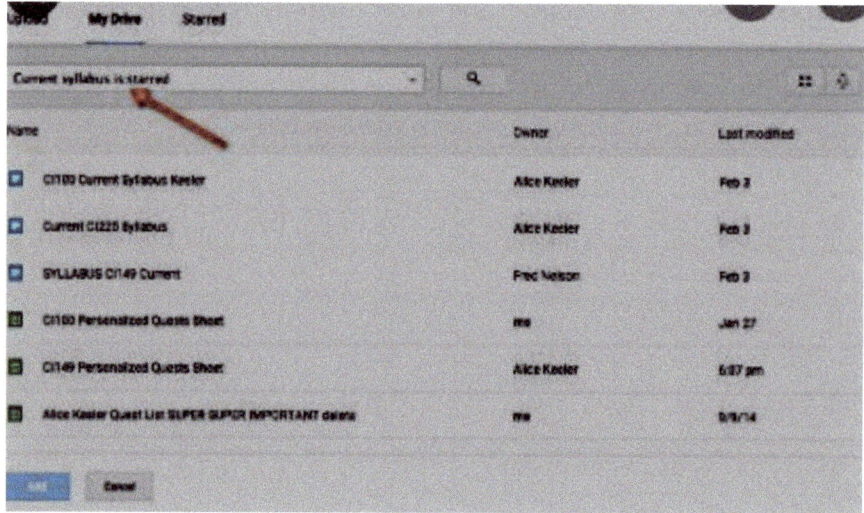

Step 1

You can post YouTube videos on the stream page or on any class assignment that you want, to do it on the stream page, you'd go through the motions of making a post, and you will see that there is an "Add" button if you go ahead and click on that it brings up four options, including a YouTube option. So the easiest way to add a YouTube video is by clicking that option.

Step 2

So go ahead and click on that and it brings a search up there that populates videos through YouTube, so you are actually on a YouTube search bar. So go ahead and type in any search that you want.

Step 3

So once you've populated your results and hit the search bar enter, you'll see all of the YouTube videos come up, and it's just one long stream of them. It seems to be endless, and it'll just keep loading more and more and more, so you can find your video that way. Go ahead and click on whatever one you want to add to the post. And when you click on it once you see that, there's a blue rectangle that encapsulates the entire selection. Then down at the bottom, you now have a blue Add button that's also rectangular that we could go ahead and hit. And notice, it is now added as an attachment to this post.

Step 4

Now, this process of adding a YouTube video is identical, no matter where you're adding it in Google Classroom, it's identical if you're in the classwork page and you click on Create. And then you click on assignment. You'll see the same type of hub, where you can put in all of your assignment details. Now, towards the bottom underneath the instructions, there, you got that paperclip again with the Add button, go ahead and click on that. You have the same four options that you did on the announcement page. You can click on the YouTube page, and it brings up that same hub, and you repeat the process.

Step 5

So as you can see, it's just as simple as adding it in your search bar, click on whatever video you want, and then click on Add. And it adds it straight to that assignment, and then you can post the assignment as you normally would, and the students can now have access to that video. Now, if for some reason you cannot find your video by searching it through this YouTube hub. But you already have the video you want on YouTube's website. Then you can grab the link and then search it and put it right in there.

Step 6

So if you're on YouTube's website, and you're on a video that you want, you can look straight up at the top of the URL address bar. You can go ahead and copy that and then go back to your classroom page. When you're in this hub to search for YouTube, instead of searching for a keyword, you are going to search that exact link and hit the search bar. You're going to see that precise video shows up. Now notice there are no other videos because this is a specific link, so it is easier to do than sifting through hundreds of videos, so you just go ahead and click "Add."

Attaching a File from Your Computer

Step 1

To add in your files from your computer or device to an assignment or select an assignment. And in the top right-hand corner, you will select "Add" or "Create", and from the list, you will choose File.

Step 2

From there, you can click Select File from the device. And then, you can navigate to your document and find where you have saved your files.

Step 3

You will select documents you would like to upload and select the "open", and you will add the file in. And then, from the bottom left-hand corner, you will select the upload button. You can repeat this process for each type of file that you would like to upload.

Step 4

Once you have finished adding your files, be sure to click "add-in", and it will ask for confirmation that you want to attach your files, click add-in, and your teacher will now be able to see your attached files.

Attaching a Link

Step 1

Begin by opening up your classroom account and then open the assignment. Once you click on the assignment open button, you'll be in the full screen, click the button that says "Add" and you'll be allowed to add a link right there.

Step 2

When it is clicked on, there's a place to put your URL, to get the URL, go to whatever it is that you need to add the link, click the Share button or the upload button, whatever it is, Ctrl C will copy that link, go back to your classroom upload page and type Ctrl V, Click "Add a link." And then, when you're finished with it and ready to return, simply click the return button.

Attaching Files from Google Drive

Step 1

First, create a material and just give it a name.

Step 2

And then go into your Google Drive. Get the folder by just clicking on your drive. So once you see the materials folder click on it, and the blue is highlighted, but the add is greyed out, and you click into it.

Step 3

If you can't add the folder using the drive. Here's what you have to do. You have to go back into that folder, and you need to get a shareable link. (That is sent to anyone with the link in your district can view) and then we go back to the classroom, so you are not going to use Google Drive you are going to use a link.

Step 4

And so then you add the link you hit paste. You hit "Add a link." And now, that it's in there, you can just click Post, and it will be posted as a material. And there it is. And that's how you do it, so you take the folder, you get the shareable link. And then, you set it to the only view, only let people view, and then its created.

Analyze and Manage Students' Answers and Scores

Then we have grading. Grading is a focal part of helping students understand and learn better, but it is a bit different with Google Classroom. Here, you'll learn all about how to easily and effectively grade work in Google Classroom.

To Begin

First, the students need to turn in the work, so once it's turned in; it's time for you to start qualifying. You need to log in and click the stream lab. You then can, if it isn't displayed already, check the assignments that are already there. You can check to see who is done and who is not done. Choose one that's above done, and from there, you'll have an expanded list on who has turned their assignment in.

From there, you should then click on the name of the student to see their assignment. If they have an attachment there, you simply click the attachment and you're given the appropriate Google app with the assignment that's on it.

At this point, you can start grading and can add comments and the like from there, too.

Commenting on Grades

At this point, you're then opening it up in Drive and can start to comment, too. If you're a teacher who likes to grade with a red pen, for example, you can essentially go to the text button, change the color to red, and then comment. But it's a bit easier this time around. If you want to, you can use the feature that allows you to comment to give the appropriate feedback. To do this, you highlight what you're about to comment on, and then choose the option to insert, and then make your comment. When you've finished typing in what you need, save it, and it will be saved completely for the student. You can mark up the assignment as needed, or even leave positive comments if there is something the student did well with.

Now from there, you can go to the class-work tab, then the assignment name, and then view it. If you haven't changed the value of the point system yet, you can always change it. From there, choose the student file that you've finished, and enter the name, then the grade. It is then ready to be returned to the student for review.

However, if you want to change the grade itself, you can go to the assignment that the student has and then enter the grade. You can also return them ungraded as needed. Remember that the changes to the grade only affect those not yet returned, and original ones have the same grade as before.

Returning Assignments to Students

Assignments can be returned to the student at this point by pressing the box that says, return. You can return it before it's recorded. When it's recorded, it's done with. You can then press the option to return the assignment.

Now, if you have additional feedback, it'll give you an option to do that too. Again, it's ultimately up to you. If you have no feedback, don't worry about that part. If you do have more feedback, then throw it in quickly before it's completely returned to the student. Remember, it's better to be a bit overboard with grading if you have feedback that will help students become better, and to help them understand the subject at hand.

Tips on This

The first tip is that when you're making assignments, don't use MS Word. Instead, use the GoogleDrive apps, since they are completely integrated with the classroom system. If you do use a Microsoft Word file, the student will have to download the files once more, upload them again, and then attach them. You will also need to put in the extra work with downloading and reattaching, and it's just a lot of extra fluff that you don't need. The Drive files are there for a reason, and they're super easy to create. However, if you export a sheet from MS Word to drive, it works the same way, making it easy for everyone.

Another helpful tip is that you can actually use shortcuts to add comments. You can use control alt and then M to put comments into a document on Google Docs. You can then press the enter key to close the comment, and control plus W to actually close the document itself. You can also do feedback by choosing the name of the students, and then looking at the options to see what they've submitted. You can go to the add private comment section on this, and also enter grades for students. You can't get a grade book with Google Classroom; just notify students of the grades.

With Google Classroom, the key way to ensure that you're getting feedback to students quickly is through adding the mobile app. It allows you to add comments to various projects, and answer questions on grades. Plus, it's integral if you want to make sure whether a student got the assignment or not.

Another important tip is to utilize the form templates to help with grades. The form template can be used to make a sheet with the names, and a checklist of various elements, including what they're missing, homework points, and other elements. By carrying this around you can also grade the students, and it's good to have if you want to check out whether or not they have their homework done.

Another really cool tip is that if you want to make your grading faster, it is possible to use shorthand. Google Docs knows immediately what you're saying, so if you use shorthand, and you type in the letters "WC" it will automatically change this to word choice, which will communicate to the students that it's a bad word choice. It makes your life so much easier, especially if you're going through grading multiple papers.

It's also important for teachers to remember that when you give an assignment back, you can't edit it anymore. That means that if you need to edit anything else, they will need to resubmit it. You can notify them, and they can look at it, and if they resubmit it with changes, and you edit it and it's all good, you can also edit the grades by looking at the grade, pressing it, and then choosing the option to update the grade.

Exporting Grades

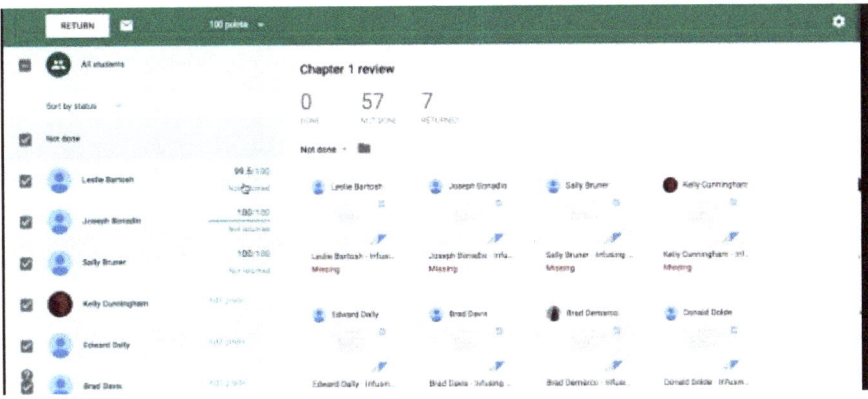

Finally, let's talk about exporting grades. Grades can be exported, and they're used to make sure that you have a place for all of them. Remember, this just displays the grades and isn't a biodegrade, but if you want to help export them so that they're all written down, you can. Lots of teachers like to export them into a .csv file, or through Google Sheets. With Sheets, you can create an average for the class and students, and set up arrows from one grade box to another one, which makes it faster.

To export these grades into sheets, you go to Google Classroom, choose your class, and then the assignments. From there, go to settings and choose the option to copy all of the grades into Google Sheets. An automatic spreadsheet will then be created on the drive folder, allowing you to see all the grades. Currently, you can only export these on the desktop version of the classroom, not on a mobile version or via the app.

Copy all grades to Google Sheets

Download all grades as CSV

Download these grades as CSV

Now, if you export these grades to a CSV file, you'll be able to have all of the grades in one place. This is good if you're trying to keep every single grade in one place, and if you want to print them out. To do this, you go to Google Classroom again, click on the class you want, then go to settings once again and choose an option, either to download the assignment grades for that one only or to download all of the assignment and question grades. For the first, you choose: download these grades as a CSV, and for the assignment and question grades, you choose: download all grades as a CSV. From there, you can find them in your downloads folder and can bring them up on your word processing device accordingly.

For many teachers, the element of grading is made so much easier with Google Classroom, since they can easily create the environment that they want and set it all set up so that students can access it easily. It's easy and simple to achieve, so you'll be able to create the best and easiest classroom experience that they can possibly have.

Google Classroom Strategies to Start Using Today

Reduce the carbon footprints of your class

The idea of Google Classroom is to make things easier for teachers and students alike when learning things. It takes the conventional classroom and places it on the online sphere and enables students and educators to create spreadsheets and presentations, online documents and it makes sharing and communicating easier. Creating and sharing things digitally eliminates the need for printing. Schools use a lot of papers, but utilizing Classroom enables you to remove the necessity of paper for simple things. Have an assignment? Save some trees, time, and money by creating them in Google Classroom, distributing it to your students in your Classroom.

Distribute and Collect Student's homework easily

The whole point of creating the assignments via Google Classroom is so that you can distribute it and collect the assignments quickly. Yes, you can say that you could get it done via email too. But Classroom enables all these things to be done in one place. You'll know who has sent an assignment, who have passed their deadline, and who needs more help with their work. It's all about lessening the hassle in your life.

Utilizing the feedback function

With instant access, teachers can clarify doubts, concerns, and misconceptions their students may have by providing feedback as and when students need it. As teachers, you eliminate possible issues that might arise while students are doing their assignments. This reduces the headache you might have upon receiving the assignments that don't meet the requirements. Assignments that are handed in that have issues can be immediately rectified as well, through private one-on-one feedback with the relevant student.

Create your personalized learning environment

The main benefit of Google Classroom is the freedom that it gives teachers. Very often, teachers are required to follow the national syllabus forwarded by the Department or Ministry of Education in a country. While this is rightly done for the sake of uniformity and to ensure students across the country have access to the same level of education, utilizing Classroom, on the other hand, gives teachers the freedom to add and create a different environment for learning.

Encourage real-world applications

Encourage students to submit their assignments using real-world material whether it's a series of videos or photos, a compilation of multimedia applications, using the many different apps out there to create amazing online presentations are just some of the things that students can do that will increase their learning tendencies and spark online discussions within the Classroom. This enables the students to apply and implement assignments that they have done in their real lives.

Allow shy students to participate

As teachers, we know which students are more extroverted than the other. Sometimes in conventional classroom settings, the shy kid or the kid with self-esteem issues or those that lack confidence and have problems participating in classroom activities, speaking out, or even raising their hand to answer questions. Google Classroom gives a safety barrier for students that fall into this category but allowing them to be more open with discussing and expressing themselves. As the teacher, you can also find creative ways to encourage these students to open up via game-based learning to promote trust, openness, teamwork, and collaboration.

Allow for coaching

Some students need more coaching and a little bit more explanation. If you know some students in your class that needs it, you can give them extra instructions by privately messaging them. You can always follow up with them while they are doing their assignments just to check if they are on the right track. Additionally, you can also invite another teacher to collaborate and help with coaching your students.

Interactive Activities Using Google Classroom

The more you use Google Classrooms, the more you will be able to use it in different ways than just connecting with your students and creating assignments. Google Classroom combined with other Google products such as Google Slides can deliver powerful interactive user experiences and deliver engaging and valuable content.

Teachers looking to create engaging experiences in Google Classroom can use Google Slides and other tools in the Google suite of products to create unique experiences.

Here are some exciting ways that you can use Google Classroom and Google Slides to create an engaging learning experience for your students:

Create eBooks via PDF

PDF files are so versatile, and you can open them in any kind of device. Want to distribute information only for read-only purposes? Create a PDF! You can use Google Docs or even Google Slides for this purpose and then save it as a PDF document before sending it out to your classroom.

Create a slide deck book

Make your textbooks paperless too, not just assignments. Teachers can derive engaging and interactive content from the web and include it in the slide deck books, upload it to the Google Classroom, and allow your students to access them. Make sure to keep it as read-only.

Play Jeopardy

This method has been used in plenty of Google Classroom and the idea was created by Eric Curts, a Google Certified Innovator, created this template that you can copy into your own Google Drive to customize with your questions and answers. Scores can be kept on another slide that only you can control.

Create Game-Show Style Review Games

Another creative teacher came up with a Google Slide of 'Who Wants to Be a Millionaire?' The template allows you to add in your questions and get students to enter the answers in the text box. Again, you keep the score!

Use Animation

Did you know you can create animations in your Google Slide and share it in Classroom? You can also encourage your students to create an animation to explain their assignments. This is making them push boundaries and think out of the box.

Create stories sand adventures

Using Google Slides and uploading them to Google Classroom to tell a story. Turn a question into a story and teach your students to create an adventure to describe their decision for the outcome of the character in their story. The stories can be a certain path that the students have chosen for the character of a story that explains the process of finding a solution.

Using Flash Cards

Flashcards are great ways to increase the ability to understand a subject or topic. Do you want to create an interactive session on Google Classroom using flashcards? You can start by utilizing Google Sheets which gives you a graphic display of words and questions and then to reveal the answers, all you need to do is click. Compared to paper flashcards, these digital flashcards allow you to easily change the questions, colors as well as the answers of the cards depending on what you are teaching the class. Digital flashcards also are an interactive presentation method that is guaranteed to engage your Classroom and bring about a new way of teaching using Google Classroom's digital space.

Make vocabulary lessons, geography lessons, and even history lessons fun and entertaining with digital flashcards.

Host an online viewing party

Get your students to connect to Classroom at a pre-determined date and time when there is a noteworthy performance, play, or even movie that is related to the subjects you are teaching in your class. Let them view the video together and also interact with them by adding questions to your Google Classroom and allowing your students to reply to you in real-time. This way, you can see assess them on their reflections, level of understanding, and their observations. You can also give your interpretation of the scene and explain it again to students who do not quite understand.

There is no limit to what a teacher can do with Google Classroom and the entire Google Suite of apps whether its Google Slides or Google Calendar or even Google Maps. The only thing you would need is creativity and the desire to give your student a different experience when using Google Classroom.

Example of Google Form Uses

CLASSROOM MANAGEMENT & DAY-TO-DAY ACTIVITIES	LESSONS, ASSESSMENT & REFLECTION
Sign-up Sheets	Interest Surveys
Class Information Management	Learner Self-Monitoring Reports
Nomination Forms	Reading Journals
Make-up Request Form for Parents	Learning Logs
Teacher/Colleague Observation	Quizzes & Tests
Attendance Check-ins	Writing Prompts

Annotated Bibliography Collection	Observation Rubrics
Assignment Submitting (Link)	Instant Feedback on Lessons or Instruction
Snack Sign-up Forms	Student-to-Student Data Collection Projects
Scheduling Parent-Teacher Conference	Data Collection for Experiments
Volunteer Sign-up	Peer Feedback
Department & School-wide Surveys	K-W-L for Assessing Prior Knowledge
Technology Issues Reporting Tool	Reading Record
Getting to Know You Class Survey	Project Progress Form
Dialoging with Parents	Observation Tools (checklists; anecdotal notes, etc.)
Placing Orders for Fundraisers	Questioning
Parental Feedback	Beginning-of-Year Technology Skills Survey
Discipline Referrals	Student-Created Choose Your Own Adventure Stories
Creating Lesson Plans	Debate Social Attitudes Form
Manage Classroom Lending Libraries	Website Evaluation Form
Parental Absence Notice	Student Note-taking
Topic Sign-up Sheets	Global Collaboration
Computer Lab Reservation Form	Exit Tickets
Grading/Observation Rubric	Reservations for Class Trips

Conclusion

The world is starting to go digital, and almost every part of our lives are being affected. From the health industry to the fashion industry, and even the educational sector, everything is becoming digitalized.

Online learning is becoming the new normal and is rapidly taking over as the accepted form of learning by educators and students all over the world. Teachers are learning to appreciate the ease of the process, and students are starting to get used to the idea of learning comfortably. The whole process of learning has become flexible, and educational theory has moved from the idea of classroom lectures, where only the teacher speaks, to one where everyone can participate. Platforms like Google Classroom further enhance the collaborative spirit of modern-day educational theory by supporting group projects, communication between teachers, students, and between students, and offering a variety of tools for collaborative learning.

Learning is now mobile and is not restricted to a physical/geographical location as it takes place on the go. The fact that students can always access these learning services and resources at any time is perhaps one of the strongest points for online learning. Forgetting or losing your book or assignments will be a thing of the past.

What makes Google Classroom distinct from other Google tools, is the teacher-student interface that is solely designed for educational activities. It allows learning to take place virtually, and it makes the learning process much easier as there are no limitations or hindrances to teacher-student communication.

The platform is also innovative, as it combines tools for instruction with tools for classroom administration like grading and participation, with tools for administrators, and it goes even further by supporting teacher-parent communication. This capability fosters a beneficial flow of communication between all parties involved in the academic well-being of the students.

Anybody can make use of Google Classroom with the right knowledge and tools at their disposal. As much as it is easy to use, proper guidance can still be beneficial about what the learning service is all about and how it can be effectively and efficiently utilized for both teacher and student. This book has, hopefully, provided you with just that extra guidance.

This book is a step-by-step guide that lets you in on the world of online learning technology, specifically by explaining the Google Classroom learning platform, and how it is becoming the future of online education.

It also offers an insight into what Google Classroom is all about and how it can be used to fulfill the objective of teaching, which is to impart knowledge to the student and make sure that whatever they learn stays with them. With this book, teachers and students are open to the benefits of using Google Classroom.

The educational sector is experiencing a profound change, and it is taking a new face with the use of online learning, which is rapidly becoming a trend and is becoming widely accepted by educators and students from all over the world. The different chapters discussed in this book are eye-opening, especially for people who have little or no idea about Google Classroom.

Google Classroom is taking over as one of the most downloaded educator learning tools and has proven to be quite useful in the learning process. With the current pandemic, it has become quite popular as a result of schools being closed. Google Classroom is being used as an alternative instead of choosing to remain idle. It has provided a way for teachers and students to continue to make academic progress despite the pandemic and school closures.

Google Classroom has some beneficial features that make it stand out, such as the feedback feature, the ability to create and distribute assignments, the ability for teachers to track and monitor student performance, and more. The presence of such useful functionalities has made it a sought after program by teachers and students alike.

Google Classroom can also be synchronized with other tools and applications. As part of this topic, we discussed the various features that are available on Google Classroom, as well as the best apps that teachers can integrate with their classrooms to enhance the process of learning and ensure that they retain their students' attention and guarantee their active participation.

Communication is important in life, and this fact is just as true when it comes to online learning, which is why the book discussed what Google Classroom interaction is all about. Google Classroom has perhaps stood out as one of the best learning tools available to educators from all over the world because of the ability for teachers to interact freely with their students in an atmosphere that is devoid of intimidation.

The teacher can receive responses and comments from their students; they can also send messages privately or as a general comment to the classroom. The platform is set up to allow two-way communication, meaning that the classrooms can grow and become true avenues for student-teacher interaction. Google has sought to recreate, as much as possible, the real classroom experience in a virtual world. Google Classroom's features also support teachers in ensuring that students also get to interact among themselves by providing features that encourage student interaction through group projects or activities, discussions, polls, and general classroom topics. Here students can interact and share ideas/information among themselves and also with the teacher. Another good thing about Google Classroom is that it allows collaboration with other classrooms from different cities and even countries, so students can communicate and interact with others. This can often lead them to create meaningful friendships, even in cyberspace.

Finally, the book provides some different tips and tricks for both teachers and students to keep up their sleeves to help them best use Google Classroom. The sheer variety of integrated applications, Google Applications, and Classroom features can boggle the mind. Without having an idea of how to effectively use these functions, it's tricky to get started. We truly hope that the lists of tips and tricks we provided will help both students and teachers make the most out of Google Classroom and adapt it to suit their unique needs.